T0191870

Birkhäuser

Modeling and Simulation in Science, Engineering and Technology

More information about this series at http://www.springer.com/series/4960

Nicola Bellomo · Abdelghani Bellouquid
Livio Gibelli · Nisrine Outada

A Quest Towards a Mathematical Theory of Living Systems

Nicola Bellomo
Department of Mathematics, Faculty of Sciences
King Abdulaziz University
Jeddah
Saudi Arabia

and

Department of Mathematical Sciences (DISMA)
Politecnico di Torino
Turin
Italy

Abdelghani Bellouquid
Ecole Nationale des Sciences Appliquées de Marrakech
Académie Hassan II Des Sciences Et Techniques, Cadi Ayyad Unviersity
Marrakech
Morocco

Livio Gibelli
School of Engineering
University of Warwick
Coventry
UK

Nisrine Outada
Mathematics and Population Dynamics Laboratory, UMMISCO, Faculty of Sciences of Semlalia of Marrakech
Cadi Ayyad University
Marrakech
Morocco

and

Jacques-Louis Lions Laboratory
Pierre et Marie Curie University
Paris 6
France

ISSN 2164-3679 ISSN 2164-3725 (electronic)
Modeling and Simulation in Science, Engineering and Technology
ISBN 978-3-319-86162-3 ISBN 978-3-319-57436-3 (eBook)
DOI 10.1007/978-3-319-57436-3

Mathematics Subject Classification (2010): 35Q20, 35Q82, 35Q91, 35Q92, 91C99

Printed on acid-free paper

This book is published under the trade name Birkhäuser
The registered company is Springer International Publishing AG
The registered company address is: Gewerbestrasse 11, 6330 Cham, Switzerland
(www.birkhauser-science.com)

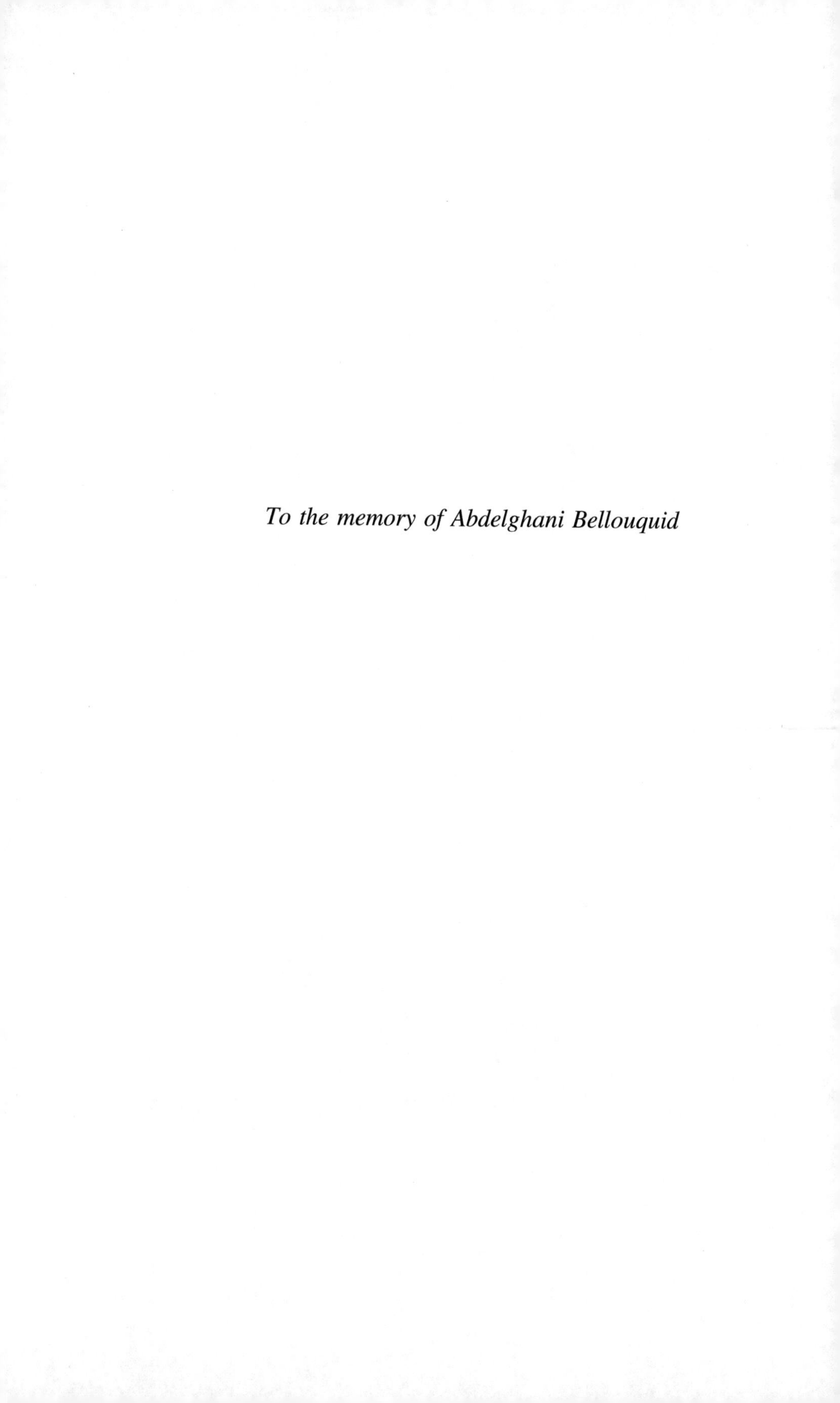

To the memory of Abdelghani Bellouquid

Preface

This book is devoted to the design of a unified mathematical approach to the modeling and analysis of large systems constituted by several interacting living entities. It is a challenging objective that needs new ideas and mathematical tools based on a deep understanding of the interplay between mathematical and life sciences. The authors do not naively claim that this is fully achieved, but simply that a useful insight and some significant results are obtained toward the said objective.

The source of the contents of this book is the research activity developed in the last 20 years, which involved several young and experienced researchers. This story started with a book edited, at the beginning of this century, by N.B. with *Mario Pulvirenti* [51], where the chapters of the book presented a variety of models of life science systems which were derived by kinetic theory methods and theoretical tools of probability theory. The contents of [51] were motivated by the belief that an important new research frontier of applied mathematics had to be launched. The basic idea was that methods of the mathematical kinetic theory and statistical mechanics ought to be developed toward the modeling of large systems in life science differently from the traditional application to the fluid dynamics of large systems of classical particles.

Here, particles are living entities, from genes, cells, up to human beings. These entities are called, within the framework of mathematics, active particles. This term encompasses the idea that these particles have the ability to express special strategies generally addressing to their well-being and hence do not follow laws of classical mechanics as they can think, namely possess both an individual and a collective intelligence [84]. Due to this specific feature, interactions between particles are nonlinearly additive. In fact, the strategy developed by each particle depends on that expressed by the other particles, and in some cases develops a collective intelligence of the whole viewed as a swarm. Moreover, it often happens that all these events occur in a nonlinear manner.

An important conceptual contribution to describing interactions within an evolutive mathematical framework is offered by the theory of evolutive games [186, 189]. Once suitable models of the dynamics at the scale of individuals have been derived, methods of the kinetic theory suggest to describe the overall system by a probability distribution over the microstate of the particles, while a balance of the number of particles within the elementary volume of the space of the microstates provides the time and space dynamics of the said distribution, viewed as a dependent variable. Quantities at the macroscale are useful in several applications, and these can be obtained from averaged moments of the dependent variable.

The hallmarks that have been presented above are somehow analogous to those proposed in the book [26], where the approach, however, was limited to linear interactions. Therefore, this present monograph provides, in the authors' belief, a far more advanced approach, definitely closer to physical reality. Moreover, an additional feature is the search of a link between mutations and selections from post-Darwinist theories [173, 174] to game theory and evolution. Indeed, applications, based on very recent papers proposed by several researchers, have been selected for physical systems, where nonlinearities appear to play an important role in the dynamics. Special attention is paid to the onset of a rare, not predictable event, called black swan according to the definition offered by Taleb [230].

The contents of the book are presented at the end of the first chapter after some general speculations on the complexity of living systems and on conceivable paths that mathematics can look for an effective interplay with their interpretation. Some statements can possibly contribute to understanding the conceptual approach and the personal style of presentation:

- The study of models, corresponding to a number of case studies developed in the research activity of the author and coworkers, motivated the derivation of mathematical structures, which have the ability to capture the most important complexity features. This formal framework can play the role of paradigms in the derivation of specific models, where the lack of a background field theory creates a huge conceptual difficulty very hard to tackle.
- Each chapter is concluded by a critical analysis, proposed with two goals: focusing on the developments needed for improving the efficacy of the proposed methods and envisaging further applications, possibly in fields different from those treated in this book. Applications cover a broad range of fields, including biology, social sciences, and applied sciences in general. The common feature of all these applications is a mathematical approach, where all of them are viewed as living, hence complex, systems.
- The authors of this book do not naively claim that the final objective of providing a mathematical theory of living systems has been fully achieved. It is simply claimed that a contribution to this challenging and fascinating research field is proposed and brought to the attention of future generations of applied mathematicians.

Finally, I wish to mention that this book represents what has been achieved until now. Hopefully, new results can be obtained in future activities. However, I decided to write a book, in collaboration with *Abdelghani Bellouquid*, *Livio Gibelli*, and *Nisrine Outada*, according to the feeling that defining the state of the art at this stage is a necessary step to look forward. *Abdelghani*, *Livio*, and *Nisrine* were kind enough to allow me to write this Preface, as my experience in the field was developed in a longer (not deeper) lapse of time. Therefore, I have stories to tell, but mainly persons to thank.

I mentioned that many results have been achieved by various authors. Among them the coworkers are very many and I will not mention them explicitly, as they appear in the bibliography. However, I would like to acknowledge the contribution of some scientists who have motivated the activity developed in this book.

The first hint is from *Helmut Neunzert*, who is arguably the first to understand that a natural development of the mathematical kinetic theory needed to be addressed to systems far from that of molecular fluids. Namely, the pioneer ideas on vehicular traffic by Prigogine should have been applied, according to his hint, also to biology and applied sciences. He organized a fruitful, small workshop in Kaiserslautern, where discussions, critical analysis, and hints left a deep trace in my mind.

Subsequently I met *Wolfgang Alt*, who also had the feeling that methods of kinetic theory and statistical mechanics in general could find an interesting area of application in biology. He invited me to an Oberwolfach workshop devoted to mathematical biology, although I had never made, as a mathematician, a contribution to the specific field of the meeting. I was a sort of a guest scientist, who was lucky to have met, on that occasion, *Lee Segel*. His pragmatic way of developing research activity opened my eyes and convinced me to initiate a twenty-year activity, which is still going on and looks forward.

However, I still wish to mention three more lucky events. The first one is the collaboration with *Guido Forni*, an outstanding immunologist who helped me to understand the complex and multiscale essence of biology and of the immune competition in particular. Indeed, my first contributions are on the applications of mathematics to the immune competition. Recently, I met *Constantine Dafermos* in Rome, who strongly encouraged me to write this book to leave a trace on the interplay between mathematics and life. Finally, I had the pleasure to listen the opening lecture of *Giovanni Jona Lasinio* at the 2012 meeting of the Italian Mathematical Union. I am proud to state that I do share with him the idea that evolution is a key feature of all living systems and that mathematics should take into account this specific feature.

My special thanks go to Abdelghani Bellouquid. He has been a precious coworker for me and several colleagues. For my family, he has been one of us. I have to use the past tense, as he passed away when the book was reaching the end of the authors' efforts. My family and I, the two other authors of this book, and many others will never forget him.

The approach presented in this book was certainly challenging and certainly on the border of my knowledge and ability. In many cases, I have been alone with my thoughts and speculations. However, my scientific friends know that I have never been really alone, as my wife *Fiorella* was always close to me. Without her, this book would have simply been a wish.

Turin, Italy
January 2017

Nicola Bellomo

Contents

Chapter 1
On the "Complex" Interplay Between Mathematics and Living Systems

1.1 Introduction

This book aims at understanding the complex interplay between mathematical sciences and the dynamics of living systems focusing on the development of mathematical tools suitable to study the dynamics in time and space of the latter.

This objective is strongly motivated by the scientific community according to the widely shared idea that one of the great scientific targets of this century is the attempt to link the rigorous approach of mathematics, and of the methods of hard sciences in general, to the study of living, hence complex, systems. Therefore, it is expected that a great deal of intellectual energies will be devoted to this fascinating objective which can be pursued, may be at least partially achieved, by inventing new mathematical methods and theories. This means going through the several theoretical sources of mathematics can offer to look for new conceptual ideas.

The main field of application of the mathematical modeling of living systems is biology, but more recently a general focus has been addressed to a large variety of fields such as opinion formation, political competition, social sciences in general, collective dynamics of systems of living entities such as crowds and swarms, financial markets, and various others.

It is not an easy task due to the lack of background field theories as put in evidence by several authors, among others, Herrero [150], May [172], and Reed [205]. The failure of several attempts is often due to the straightforward, sometimes uncritical, application to living systems of classical methods which are valid only for the inert matter. Indeed, the lack of success of several attempts generated the pessimistic statements by Wigner [243].

The growing interest in the interplay between mathematics and biology has generated several interesting hints addressed to mathematicians and physicists. An important example is offered by the Nobel Laureate Hartwell [137] who, focusing on biological systems, indicates some important features which distinguish living systems from the inert matter. In more detail:

N. Bellomo et al., *A Quest Towards a Mathematical Theory of Living Systems*, Modeling and Simulation in Science, Engineering and Technology,
DOI 10.1007/978-3-319-57436-3_1

Biological systems are very different from the physical or chemical systems analyzed by statistical mechanics or hydrodynamics. Statistical mechanics typically deals with systems containing many copies of a few interacting components, whereas cells contain from millions to a few copies of each of thousands of different components, each with very specific interactions.

Although living systems obey the laws of physics and chemistry, the notion of function or purpose differentiates biology from other natural sciences. Organisms exist to reproduce, whereas, outside religious belief, rocks and stars have no purpose.

Selection for function has produced the living cell, with a unique set of properties which distinguish it from inanimate systems of interacting molecules. Cells exist far from thermal equilibrium by harvesting energy from their environment.

Therefore, we can learn from biologists, specifically from the first paragraph of the above citation, that living systems are characterized by a structural complexity which, if not properly reduced, can generate mathematical problems that cannot be technically handled by analytic and/or computational methods, for instance due to an excessive number of coupled equations. Thus, reducing complexity is a structural need of the mathematical approach which, however, should not loose the most relevant features of each system under consideration.

This issue is further stressed in [137] by the second and third paragraphs of the citation, which suggest that mathematical approaches need to take into account the ability of living systems to develop specific strategies induced, not only by interactions with other entities and with the outer environment, but also by the ability to adapt themselves to evolving conditions as a result of reasoning and learning abilities.

Hartwell’s sentences anticipate a concept to be taken into account within any mathematical approach, namely the presence of an evolutive and selective dynamics, which can be referred, in a broad sense, to a Darwinian-type dynamics [174].

There is a long story of philosophical contributions to this topic, whose essence has been put in evidence in the following sentence by Jona Lasinio [160]:

Life represents an advanced stage of an evolutive and selective process. It seems to me difficult understanding living entities without considering their historical evolution.

Population dynamics is based on a rather primitive mathematical theory; on the other hand, it should explain the emergence of individual living entities by selection.

Moreover, the conclusive question is:

Is life an emerging property of matter?

The original article is in Italian; hence, both sentences have to be viewed as a free translation of the authors of this book.

These reasonings put clearly in evidence the evolutive feature of living systems. Therefore, mathematicians are required to invent new mathematical tools suitable to describe the dynamics related to Darwinian and post-Darwinian theories of evolution.

However, the interest to study the mathematical properties of complex system is not limited to biology as complexity occurs almost everywhere in societies whose dynamics receives important inputs from human behaviors. A broad variety of environments, where complexity appears, is reported in the collection of papers [18]. Therein, it is also well documented how specific studies can be addressed to contribute to the well-being of our society [141]. Some of these topics will be treated in the next chapters by the approach proposed in this book.

The interest of mathematicians in new structures is a fascinating field of research in mathematical sciences. The reader can overview the paper by Gromov [132], Abel Prize Laureate, where the abstract introduces some author's ideas on this topic, while the paper analyzes a number of entropy functionals looking at their ability to retain properties of real systems.

Mathematics is about "interesting structures". What makes a structure interesting is an abundance of interesting problems; we study a structure by solving these problems.

The worlds of science, as well as of mathematics itself, is abundant with gems (germs?) of simple beautiful ideas. When and how many such an idea direct you toward beautiful mathematics?

All these reasonings indicate that the mathematical approach to the modeling of living systems requires a deep understanding of the complexity of life in its broad variety of aspects. Therefore, this first chapter presents some conceptual speculations concerning the interpretation of the main features of living, hence complex, systems. In detail:

- Section 1.2 provides a concise overview of three scientific works that represent key contributions to pursue the aim of the book. Their contribution to design the sequential steps of the mathematical approach is definitely important.

- Section 1.3 brings to the readers' attention five key questions related to the attempt of mathematical sciences to model large living systems. These questions motivate the contents of the whole book. The answer to the first question is given in this chapter, while the others will be given in the next chapters.

- Section 1.4 presents a selection of complexity features, hopefully the most relevant ones, which characterize living systems. This selection looks forward to modeling approaches, where specific models should capture the most relevant features for each specific system.

- Section 1.5 anticipates the strategy followed to pursue the development of the mathematical approach not only toward the derivation of models of living systems, but also in view of the development of a proper mathematical theory. Subsequently, starting from this strategy, the plan of the book is presented.

1.2 A Quest Through Three Scientific Contributions

Arguably, the first relevant approach toward the ambitious aim of casting some aspects of living systems into a mathematical framework is documented in the monograph by Schrödinger [220], motivated by the study of mutations (some of them also induced by external actions such as radiations) in living systems.

It is worth observing that nowadays, physicists and biologists consider the great physicist Erwin Schrödinger as the precursor of the modern molecular biology supported by genomic and post-genomic research. Moreover, he introduces the concept of multiscale modeling of biological systems, where the dynamics at the low cellular scale determines the overall dynamics of the system. This concept is nowadays the most important guideline of the interplay between hard sciences and biology, where understanding the link between the dynamics at the molecular scale of genes and the functions expressed at the level of cells is one of the important, may be the most important, objective of the development of biomathematical theories.

It is plain that dealing with the complexity of living systems leads naturally and rapidly to biology. However, it can be claimed that living (complex) systems show common features which appear in a large variety of fields for any system related to the living matter. For instance, we can move to a completely different field through the contribution by Prigogine and Herman [202] to the modeling of vehicular traffic on roads. Actually, their approach has been pragmatic and simply aimed at providing mathematical models of the dynamics of vehicles on roads. However, their book contains perspective ideas and mathematical tools, which can potentially model also other living systems of the real world.

The interpretation that Prigogine's model is a technical modification of the Boltzmann equation is definitely unfair, as his book [202] contains new general ideas on the modeling of interactions between vehicles by tools of probability theory and on the modeling of heterogeneous behaviors. In fact, this book anticipates issues that now are developed in various fields of applications such as the dynamics of crowds and swarms. The great conceptual novelty is that the overall state of a large population is described by a probability distribution over the microscale state of interacting entities and that the output of interactions is delivered by probability rules rather than by the deterministic causality principle.

An additional pioneer paper, worth to be mentioned, has been published by Jäger and Segel [159], where the social dynamics of certain species of insects (bumble bees) is studied. This paper introduces the concept of social internal state (later we will refer to it as *activity*) to model the microstate of insects, called *dominance*, which is the ability of an insect to impose a behavioral strategy to the others. Social interactions split the society into dominant and dominated. In addition, this paper considers as an important feature that the overall state of the system is described by the probability distribution over the microstate of the interacting insects, while the output of interactions is described by probability rules.

This paper [159] exerted an important influence over the research activity devoted to the modeling of social systems. Some of the pioneer ideas of [159] have contributed, as we shall see, to the hallmarks of the mathematical methods proposed in this book.

The just presented excursus shows how mathematical approaches used to model the dynamics of systems of the inert matter generally fail when they are applied to living systems due to the lack of a field theory [172]. Therefore, new methods need to be developed. Consequently, almost the whole content of this book is limited to this specific objective to be followed by modeling applications. Only in the last chapter the concept of mathematical theory is introduced and some perspective ideas are presented to show how to move from modeling issues to a self-consistent theory.

1.3 Five Key Questions

Often a research program is initiated by a number of key questions that any researcher poses to himself/herself. The program should pursue well-defined objectives including an answer to the said queries. Therefore, five key questions are brought to the self-attention of the authors as well as to that of the reader. An answer to all of them will be given all along the various chapters of this book starting from the next section of this chapter.

1. ***Which are the most relevant common complexity features of living systems?***
2. ***Can appropriate mathematical structures be derived to capture the main features of living systems?***
3. ***How can mathematical models be referred to the mathematical structures deemed to depict complexity features of living systems?***
4. ***Models offer a predictive ability, but how can they be validated? In addition, can rare events, namely the so-called Black Swans [230], be predicted by them?***
5. ***Which are the conceptual paths which might lead to a mathematical theory of living systems?***

As mentioned already, an answer will be given to each of these questions. The first answer is specifically given in this first chapter, while the others are given in the next chapters. Indeed, the authors' answer to the said questions defines the overall strategy of this monograph, which aims at contributing to the complex interplay between mathematics and life sciences.

The said answers cannot yet be found in the literature; hence, putting questions and providing some preliminary answers can contribute to the search of appropriate mathematical tools, and, possibly, a new mathematical theory.

1.4 Complexity Features of Living Systems

Let us now consider a large system constituted by several living entities which interact among themselves and with the outer environment. The first step of a quest along the path introduced in Section 1.1 consists in understanding the main features that characterize the complexity of living systems. So as possibly reduce complexity and cast the said features into mathematical equations without, however, losing their descriptive ability.

The next chapters also show how this effort can contribute to understand how empirical data can be used to validate models. Therefore, this section attempts to give an answer to the first one of the five key questions proposed in the preceding section, namely:

Which are the most relevant common complexity features of living systems?

Bearing all above in mind, ten key features are selected, among several conceivable ones, without claiming that it is an exhaustive choice. Indeed, it is a personal interpretation proposed according to the authors' bias to be regarded as a first step toward the development of a modeling approach. It is worth stressing that not all complexity features apply to all systems. The study of each of them should put in evidence the specific ones which need to be properly retained by models.

Each feature is mainly referred to biological systems, where all of them are generally present. However, the next chapters show that the said features appear, in a minor or greater extent, in different fields too, such as social, economical, and engineering sciences.

1. *Large number of components*: Complexity in living systems is induced by a *large number of variables*, which are needed to describe their overall state. Therefore, the number of equations necessary to the modeling approach might be too large to be practically treated, as already stated in Section 1.1.

These reasonings, as we have seen, have been proposed by Hartwell [137], who observes that biological systems are different from the physical systems of the inert matter generally constituted of a few interacting components, since multicellular systems contain even thousands of different components, each with specific biosystems. This observation, which cannot be put in doubt, obliges applied mathematicians to tackle the problem of reducing the said complexity.

2. *Ability to express a strategy*: Living entities have the ability to develop specific *strategies* related to their *organization ability* depending on the state of the entities in their surrounding environment. These can be expressed without the application of any principle imposed by the outer environment, although interaction with the environment has an influence on the said ability. A strategy can be defined "rational" when it is properly finalized to a concept (individual or collective) well-being, while it is "irrational" when, even if motivated by a well-being purpose, leads to a disaster. Of course, a challenging topic consists in understanding the difference between individual and collective well-being.

3. *Heterogeneity*: The ability, defined in (2), is *heterogeneously distributed*, as individual behaviors in living systems, composed by many entities, can differ from an entity to another. In biology, this is related to different phenotype expressions generated by the same genotype. In some cases, these expressions induce genetic

diseases [242] or different evolution of epidemics [92, 96]. The book by Nowak [186] (Chapter 5) puts well in evidence the role of heterogeneity in the physics of evolution and provides hallmarks to deal with it by theoretical tools of evolutive game theory.

4. *Behavioral stochastic rules*: Living entities, at each interaction, *play a game* with an output that depends on their strategy often related to surviving and adaptation ability, namely to an individual or collective search of fitness [52]. The output of the game is not, generally, deterministic even when a causality principle is identified. This dynamics is also related to the fact that agents receive a feedback from their environments, which modifies the strategy they express adapting it to the mutated environmental conditions [173, 174]. Moreover, interactions can modify the outer environment; hence, the rules governing the dynamics of interactions evolve in time.

5. *Nonlinear interactions*: Interactions are *nonlinearly additive* and involve immediate neighbors, but in some cases also distant entities. In fact, living systems have the ability to communicate and may possibly choose different interaction paths. In some cases, the topological distribution of a fixed number of neighbors can play a prominent role in the development of strategy and interactions. Namely, living entities interact, in certain physical conditions, with a fixed number of entities rather than with all those in their visibility domain. Non-additivity is a consequence of their ability to develop a strategy by which each individual selects the received information and organizes it in a nonlinear manner.

6. *Learning ability*: Living systems have the *ability to learn from past experience*. Learning succeeds in modifying what we have called ability to express a strategy; as a consequence, the said strategy and the rules of mutual interactions evolve in time due to inputs received from the outside. In addition, living systems succeed in adapting themselves to the changing in-time environmental conditions [174]. Learning dynamics appears, as a possible example, in the immune defense [83], when immune cells transfer the so-called *innate immunity* into the *acquired immunity*. This feature is even more explicit in the case of social systems, where individuals and groups of interest have the ability to develop, from past experience, a proper learning dynamics.

This dynamics is not simply individual based, as a living entity can learn not only by pair interactions, but also by interactions with a whole population. Then, the individual can transfer the knowledge to her/his population.

7. *Darwinian mutations and selection*: In each birth process, mutations might appear and bring new genetic variants into populations. Natural selection then screens them: by reducing the frequency of "bad" (relatively unfit) variants and increasing the frequency of "good" (relatively fit) ones. Geneticists have shown that the more a given type within a population is filter, the more rapidly it increases in frequency. Actually, natural selection seems to have a very sharp eye as it is able, in a population of a million individuals, to operate on fitness differences as small as one part in a million [188, 214, 215, 226].

This feature also appears in the study of social systems. In fact, evolutionary processes can generate new groups of interest, some of which will increase their presence as they are better adapted to the social and economical environment, while some others will disappear for the opposite reason. Therefore, post-Darwinian theories draw a parallel between sociology and developmental biology.

8. *Multiscale aspects*: The mathematical approach always needs *multiscale methods*. For instance, in biology, the dynamics at the molecular (genetic) level determines the ability of cells to express specific functions. Moreover, the structure of macroscopic tissues depends on the dynamics at the lower scales, namely molecular and cellular. In general, we can state that one only observation and representation scale is not sufficient to describe the dynamics of living systems.

This feature clearly appears also in social systems, where individuals can aggregate into groups of interest, whose collective behavior is induced by individual-based interactions and dynamics. The main problem consists in referring analytically the dynamics of the interacting entities, referred later as the *dynamics at microscale*, to the observed collective behaviors which can be described, at the so-called *macroscale*, by partial differential equations.

9. *Time is a key variable*: There exists a timescale of observation and modeling of living systems within which evolutionary events occur that generally change substantially the strategy expressed by individuals. Such a timescale, in biology, can be very short for cellular systems and very long for vertebrates. In some cases, such as the generation of daughters from mother cells, new cell phenotypes can originate from random mistakes during replication.

In the case of vertebrates, the event corresponds to individuals with an phenotype made of a composition of different features. In general, the mathematical approach should set the observation and modeling lapse of time related to the specific objective of the investigation and take into account that time and space scales might substantially differ.

10. *Emerging behaviors*: Large living systems *show collective emerging behaviors that cannot directly related to the dynamics of a few entities*, but are often generated by a kind of swarming intelligence that collectively involves all the interacting individuals [67, 84].

Generally, emerging behaviors are reproduced at a qualitative level given certain input conditions, though quantitative matches with are rarely obtained by observations. In fact, small changes in the input conditions generate large deviations. Heterogeneity of individual strategies, learning ability, and interactions with the outer environment largely influence such phenomena. In some cases, large deviations break out the macroscopic (qualitative) characteristics of the dynamics, whence substantial modifications can be observed. These deviations are requisites for a rare event, say a *Black Swan*, to develop as we shall see in Chapter 3.

Additional features can be certainly identified in real systems. However, we feel comfortable to state that this selection already retains the essential elements toward a rewarding modeling approach, which also needs reducing complexity by focusing on those aspects that effectively have a role in the dynamics of each specific system under consideration. In some cases, only a part of the said ten features needs to be taken into account. Moreover, it can be noticed that exemplifications have been referred to biological systems, where all ten features can be identified. However, the applications presented in the following show how these features also characterize different fields of life sciences. Various examples are given and critically analyzed in the next chapters.

Bearing all above in mind, let us now present some speculations on the nonlinearity features of interactions that can possibly contribute to refining the quest toward a mathematical approach. According to the style of this present chapter, only a qualitative analysis is proposed, while formal issues are left to the next ones.

Two specific issues are selected, among various conceivable ones, according to the author's bias:

- Let us consider the conjecture, proposed in [19], concerning interactions in swarms, which states that individuals in a swarm, viewed as active particles, interact only with a fixed number of other particles and not with all those which are in their visibility domain. Empirical data [80] support this conjecture. Therefore, applied mathematicians are required to tackle a complex nonlinearity feature, which implies non-locality of interactions depending on the distribution of particles (see [52]). This type of interactions can lead to large deviations in the dynamics. Therefore, one can argue that if a particle loses part of the information, for instance when the number of other particles in the visibility zone is below the aforesaid critical number, then the flocking action of the collective behavior can be weakened and particles do not operate in the best way for their well-being. A situation like this can occur during the attack of a predator, who tries to isolate a single individual from the swarm [52].
- Another important source of nonlinearity is the lack of additivity in interactions. Namely, the action of a number of active particles over another one is not the sum of all single actions. For instance, while two individuals are interacting, the presence of the third individual can substantially affect also their interaction. Therefore, multiple interactions should be dealt with at once, since each individual interacts not only with any other individual, but also with any binary interaction between two other individuals.

Waiting for the technical development of mathematical tools, in Chapter 3, with applications in Chapters 5 and 6, we will take the liberty to develop some reasoning, without using the cold language of mathematics, on the story of Kate, Jules and Jim, in the movie "Jules and Jim," directed by François Truffaut, based on the novel of the same title by Henri-Pierre Roché, and played by Jeanne Moreau (Cathérine), Oskar Werner (Jules), and Henri Serre (Jim). Quoting from [34]:

> **Jules and Jim are two young men in Paris in the first years of the second decade of the 20th century, and Cathèrine is a young lady who attracts both of them and is attracted by them. Jules is Austrian and Jim French; they are dear friends. Jules shows his psychological attitude much more clearly than Jim, who seems at first to hide with care his real feelings under the species of a simple friendship.**

Cathèrine seems in turn to became soon very fond of both of them, but perhaps to feel a bit of attraction toward Jim. Nevertheless, she marries Jules, after he proposes to her. It is plain that this is a story which involves triple interactions, as each binary interaction is constantly influenced by the presence of the third person.

Initially the three friends live in a closed environment. However, when the First World War starts, Jules is obliged to go back to Austria with Cathèrine, and the two friends find themselves fighting against each other. After the end of the war, the three friends meet again and try to live and spend some time together as before, but something is severely changed. In addition, the external environment plays a relevant role in the evolution of the story, as is clear if one notes that the system before the war can be regarded as a closed system, while it becomes open to external actions during the war. This action appears to apply a crucial effect, as the evolution after the war is very different from the first part of the story, although the system appears to be again closed.

In the last period of the story, Cathèrine falls into a deep crisis which leads her to kill herself together with Jim.

This tragic event does not seem, at the first look, predictable. Indeed, it looks like a black swan (Fig. 1.1) according to the definition by Taleb [230].

"A Black Swan" is a highly improbable event with three principal characteristics: it is unpredictable; it carries a massive impact; and, after the fact, we concoct an explanation that makes it appear less random, and more predictable, than it was.

Referring to their story, one is naturally led to ask whether a mathematical model might be constructed to describe such a complex, probably unpredictable, system. The first answer should be certainly negative. But still, applied mathematics can try to capture and describe some small subsets (particular aspects) of the system. A very preliminary attempt is given in [34], while deterministic dynamical systems have been proposed by various authors to model triple interactions, see [206] and therein cited bibliography. Our approach is naturally stochastic, while triple interactions are

Fig. 1.1 From mathematics to the black swan

only a first step toward the need of dealing with multiple interactions and, in general, with the interplay between individual and a whole population.

1.5 Rationale Toward Modeling and Plan of the Book

Let us consider systems composed of many interacting entities, where the microscale refers to individual entities, while the macroscale is related to observable quantities corresponding to the collective dynamics. Still remaining at a qualitative level, namely without taking advantage of a mathematical formalization, some guidelines that can contribute to the modeling approach can be listed as follows:

- Understanding how the complexity features of living systems can influence their dynamics.
- Deriving a general mathematical structure, consistent with the aforesaid features, with the aim of offering the conceptual framework toward the derivation of specific models.
- Derivation of models corresponding to well-defined classes of systems by implementing the said structure with suitable models of individual-based interactions, according to a detailed interpretation of the dynamics at the microscale.
- Validation of models by comparison of the predicted dynamics with empirical data.
- Analysis of the gap between derivation of models and mathematical theory.

Apparently, this approach is the classical one of applied mathematics. However, in the case of inert matter, a background field theory is available. A classical example is given by Newtonian mechanics which implements conservation equations for mass, momentum, and energy.

On the other hand, this is not the case when dealing with living systems, where no background theory supports the derivation of models. Moreover, the great heterogeneity, which characterizes living systems, induces stochastic features that cannot be hidden, nor replaced by a noise.

All reasonings developed up to this point can lead to some additional ones concerning the interpretation of empirical data, which traditionally are used to validate models in order to confirm their ability to quantitatively reproduce real situations. The said data can provide the assessment of models by identification of their parameters.

This rationale is still valid in the case of complex system; however, additional issues have to be carefully taken into account. In fact, as we have seen, the collective dynamics of complex systems is determined by interactions at the microscopic scale ruled by the strategy that interacting entities are able to express. This collective dynamics exhibits emerging behaviors as well as some aspects of the observable dynamics which may not be specifically related to complexity, such as steady conditions uniform in space that are reproduced in analogy with classical systems.

Our considerations lead to state that models should have the ability to depict both emerging behaviors far from steady cases and steady behaviors. This also implies that both behaviors should not be artificially imposed in the structure of the model, rather it should be induced by interactions at the microscale. The examples treated, after Chapter 4, provide some practical interpretation of these reasonings.

Accordingly, empirical data should be used toward the assessment of models at the microscale, and subsequently, validation of models should be obtained by investigating their ability to depict emerging behaviors. However, the process can be implemented if the modeling at the microscale is consistent with the physics of the real system and if the tuning method leads to a unique solution of the inverse problem of parameters identification.

After all above preliminaries, a description of the contents of chapters of this book can be given.

Chapter 1 has been devoted to a speculative insight into the complexity features of living systems and to a presentation of a specific strategy toward modeling. One of the main complexity features consists in the ability of interacting entities to express behavioral strategies, which are modified according to the state and strategy of other entities. Further, some very preliminary reasonings on the validation of models have been presented.

Chapter 2 provides a brief introduction to the mathematical kinetic theory of classical particles with special emphasis on the celebrated Boltzmann equation. This chapter also presents some models of interest to the specific contents of this book, namely the so-called discrete velocity Boltzmann equation, where the velocity variable is supposed to attain a finite number of speeds; the Enskog equation, which includes some features related to finite dimensions of particles; and mean field models, where interactions are ruled by attractive–repulsive potentials and are smeared in space. The last part of this chapter is devoted to a brief introduction to Monte Carlo methods, which appear to be the appropriate computational tool to obtain simulations of the various models presented in the next chapters.

Chapter 3 firstly deals with the derivation of mathematical tools for the modeling of living systems focusing on networks constituted by interconnected nodes, where individuals are supposed to be homogeneously distributed in the territory of each node. Subsequently, this chapter deals with the dynamics over space viewed as a continuous variable. The final objective consists in deriving general structures suitable to capture, as far as it is possible, the complexity features of living systems. These structures can offer the conceptual background toward the derivation of specific models related to the broad class of systems. Therefore, this chapter proposed the answer to the second key question.

Chapter 4 can be viewed as a bridge between the first chapters devoted to methodological topics and the remaining chapters devoted mainly to applications. The first section proposes a general approach to derive specific models referring to the aforementioned mathematical structures, and subsequently, the various applications treated in the second part of the book are introduced enlightening on the different features across them. The critical analysis in the last section of this chapter proposes a preliminary answer to the third and fourth key questions, while an additional speculation on the said key questions is left to the following chapters.

The presentation of the next two chapters, devoted to applications, follows a common style and consists first in a critical analysis of the existing literature, which is followed by the presentation of a specific model. The presentation include simulations proposed to enlighten the predictive ability of the model. Then, additional applications based on a similar modeling approach are briefly presented. Finally, the chapters are concluded by some theoretical speculations, which starting from the applications, look ahead to conceivable developments in new research fields.

Chapter 5 presents a first application corresponding to the general framework of Chapter 3. In detail, the modeling of social systems, focusing on dynamics of criminality in urban areas, is dealt with in various contexts. Some case studies contribute to a deeper understanding of the modeling approach and interpretation of the result of simulations.

The modeling of Darwinian dynamics in the competition between cancer and immune cells is introduced, as an additional application. This class of models differs from that of the application in the first part of the chapter by an important feature. Indeed, in the first case, the overall number of particles can be assumed to be approximately constant, while in the second case, proliferative and/or destructive phenomena play an important role. This topic is also related to a revisiting of population dynamics by hybrid systems which include both deterministic and stochastic interactions.

Chapter 6 presents an overview of modeling topics of crowd dynamics. The main difference with respect to the application proposed in Chapter 5 is that now, the dynamics is developed also over the space variable which appears both in the microscopic state of the interacting individual entities and as an independent variable of the differential equations modeling the dynamics of the system under consideration. The overall dynamics is quite complex due to the presence of walls and internal obstacles. Moreover, the strategy developed by walkers can be modified by environmental conditions, such as when panic conditions appear. Some simulations are devoted to understand the ability of models to depict equilibrium conditions referring to what is known by empirical data. These simulations also aim at validating models.

This chapter also shows how the modeling approach can be applied to model the dynamics over space of multicellular systems. Both modeling and computational topics are presented.

An additional topic, proposed in this chapter, is the modeling of swarms, where the literature in the field is rapidly growing in recent years starting from the celebrated model by Cucker and Smale [87] (see also [86]). The presentation critically analyzes how several challenging problems have not yet found a satisfactory answer, for instance, the role of leaders, collective intelligence in reaction to the presence of an external action, and many others. These applications are presented as possible speculations toward possible research activity.

Chapter 7 presents an overview of open problems concerning both modeling and analytical issues, and an important topic is the derivation of macroscopic models from the underlying description at the microscopic scale delivered by the class of kinetic equations proposed in this book. This approach generates new classes of continuum models which were not foreseen by the phenomenological derivation. The features of these models are directly related to the dynamics at the microscopic scale that contributes to a deeper interpretation, hopefully validation, of these models.

Finally, the focus goes back to a problem posed in this first chapter, namely the complex interplay between mathematics and life sciences, having in mind the ambitious goal of designing a mathematical theory of living systems, namely an answer to the fifth key question. Some speculations are brought to the attention of the reader without naively claiming that an answer is given to this challenging topic. Simply, the viewpoint of the authors is presented looking ahead to the research activity of future generations of applied mathematicians. Indeed, one of the aims of this book is to bring our personal experience to the attention of the readers and motivate research activity in the field.

Chapter 2
A Brief Introduction to the Mathematical Kinetic Theory of Classical Particles

2.1 Plan of the Chapter

The contents of this chapter are motivated by the second key question posed in Section 1.3, namely by the search of mathematical tools suitable to model living systems (specifically, we deal with systems composed of many interacting entities). The strategy to pursue this challenging objective consists in designing a general mathematical framework suitable to capture, by a differential system, the ten key complexity features presented in Section 1.4. This topic will be exhaustively treated in the next chapter. However, the classical kinetic theory represents an important reference background. Therefore, this present section provides a brief introduction to this topic.

Bearing this in mind, let us report about the methods of the kinetic theory of diluted gases and, specifically, of the celebrated Boltzmann equation. A common feature shared by the classical and the new approach is that the state of the overall system is described by a probability distribution over the state at the microscopic scale of the interacting entities, namely classical particles in the first case and active particles in the second case, where the microscopic state includes, as we shall see, an additional variable suitable to model the living features of each specific system.

The time and space dynamics of the said distribution is obtained, in both cases, by a balance of particles in the elementary volume of the space of the microscopic states. However, substantial conceptual differences can be put in evidence. In the classical kinetic theory, interactions are reversible, binary, and ruled by laws of classical mechanics, while in the kinetic theory of active particles, they are not reversible, may involve multiple entities, and do not follow, as already mentioned, rules from a field theory.

Understanding the important conceptual differences between these two approaches can contribute to a deeper insight over the new approach. Therefore, a brief introduction to the Boltzmann equation will be presented also with the aim of enlightening the said differences.

N. Bellomo et al., *A Quest Towards a Mathematical Theory of Living Systems*,
Modeling and Simulation in Science, Engineering and Technology,
DOI 10.1007/978-3-319-57436-3_2

Details of the contents of this chapter can now be given:

- Section 2.2 provides a brief introduction to the derivation of the Boltzmann equation and to its properties. This section also presents an overview of mathematical problems generated by the application of the models to the study of real flows, namely the initial value problem in unbounded domains within the framework of non-equilibrium thermodynamics and initial-boundary problems for internal or external flows.
- Section 2.3 rapidly introduces some generalizations of the Boltzmann equations, namely the so-called discrete velocity Boltzmann Equation (in short BE) to obtain a mathematical model apparently simpler than the original model, the Enskog equation, where the dimension of the particles is not neglected with the aims of modeling some dense gas effects, and Vlasov-type equation, where interactions with self-consistent force fields are accounted for.
- Section 2.4 provides a brief introduction to numerical methods which can be used to solve kinetic equations. The presentation remains at a technical level, namely for practical use, while the reader is addressed to the pertinent literature for a deeper insight into theoretical topics. Later in the book, this approach will be adapted to the specific case studies proposed in the next chapters.
- Section 2.5 proposes a critical analysis focused on the substantial differences of the physical framework, which leads to the classical kinetic theory and that of complex living systems.

It is worth stressing that the contents of this chapter is presented at a very introductory level. The reader who already possesses a basic knowledge of the mathematical kinetic theory can rapidly skip over this chapter and simply look at the last critical section. On the other hand, the reader who might think that it is exhaustive is invited to forget about this idea and go to the book by Cercignani, Illner, and Pulvirenti [82], which is the main reference for the contents of this present chapter.

2.2 Phenomenological Derivation of the Boltzmann Equation

Let us consider a system composed by spherically symmetrical classical particles modeled as point masses, where their microscopic state is given by position $\mathbf{x} \in \mathbf{R}^3$ and velocity $\mathbf{v} \in \mathbf{R}^3$, and suppose that the mean free path is large enough that only binary collisions are important; namely, the probability of multiple collisions is not significant.

The system is supposed to be set in an unbounded domain with periodic boundary conditions or with decaying density and velocity of particles at infinity.

Molecular physical theories allow to define an interaction potential of the forces exchanged by pair of particles, and hence, the exchanged positional forces can be formally written as follows:

$$\varphi_{ij} = \varphi(\mathbf{x}_i, \mathbf{x}_j), \qquad \mathbf{x}_i, \mathbf{x}_j \in \mathbf{R}^3, \quad i \neq j, \tag{2.1}$$

under the assumption that binary interactions are not influenced by the other particles.

When physical conditions make plausible all above assumptions, the dynamics of the particles is described by a large system of ordinary differential equations as follows:

$$\begin{cases} \dfrac{d\mathbf{x}_i}{dt} = \mathbf{v}_i, \\ m\dfrac{d\mathbf{v}_i}{dt} = \displaystyle\sum_{j=1}^{N} \varphi(\mathbf{x}_i, \mathbf{x}_j) + \mathbf{F}_i(\mathbf{x}_i), \end{cases} \tag{2.2}$$

where $i = 1, \ldots, N$ for a system of N equal particles subject to an external positional field $\mathbf{F}_i(\mathbf{x})$. The exchanged force φ_{ij} is repulsive for very short distances and attractive for large ones depending on the specific physical features of the interacting pairs. Further developments can take into account also a velocity dependence.

Disparate gas mixtures can be treated by the same approach, by including interactions of particles with different mass and adding new terms related to interactions of all components of the mixture.

Macroscopic quantities such as mass, linear momentum, and kinetic energy can be computed by local averages in small finite volumes. Fluctuations cannot be avoided by this technical averaging process.

An alternative to this approach consists in representing the overall system by a probability distribution and by looking for an equation suitable to describe the dynamics in time and space of the said distribution. Bearing this in mind, let us introduce the N-particles distribution function

$$f_N = f_N(t, \mathbf{x}_1, \ldots, \mathbf{x}_N, \mathbf{v}_1, \ldots, \mathbf{v}_N), \tag{2.3}$$

such that $f_N d\mathbf{x}_1 \ldots d\mathbf{x}_N d\mathbf{v}_1 \ldots d\mathbf{v}_N$ represents the probability of finding, at time t, and a particle in the elementary volume of the N-particles in the phase space and the microscopic state at time t is the vector variable defined in the elementary volume of said space of all positions and velocities.

If the number of particles is constant in time and if the dynamics follows Newton rules, a continuity equation in the phase space is satisfied. Formal calculations, related to conservation of particles, yield

$$\partial_t f_N + \sum_{i=1}^{N} \mathbf{v}_i \cdot \nabla_{\mathbf{x}} f_N + \frac{1}{m} \sum_{i=1}^{N} \nabla_{\mathbf{x}}(\mathbf{F}_i \cdot f_N) = 0, \tag{2.4}$$

where m is the mass of the particle and $\mathbf{F}_i = \mathbf{F}_i(t, \mathbf{x})$ is the force applied to each particle by the overall system as shown by Eq. (2.2).

This equation, which is known as "Liouville equation", is not practical as it involves a too large number of variables. The "great" idea by Ludwig Boltzmann has been the description of the overall state by a continuous probability distribution over the microscopic state of the particles, namely position and velocity of the *test particle* representative of the whole. Therefore, he introduced the one-particle distribution function as follows:

$$f = f(t, \mathbf{x}, \mathbf{v}) : \quad [0, T] \times \mathbf{R}^3 \times \mathbf{R}^3 \to \mathbf{R}_+, \tag{2.5}$$

such that under suitable integrability conditions $f(t, \mathbf{x}, \mathbf{v})\, d\mathbf{x}\, d\mathbf{v}$ defines the number of particles that at time t are in the elementary volume $[\mathbf{x}, \mathbf{x} + d\mathbf{x}] \times [\mathbf{v}, \mathbf{v} + d\mathbf{v}]$ space, called phase space, of the microscopic states.

Suppose now that

$$\mathbf{v}^r f(t, \mathbf{x}, \mathbf{v}) \in L_1(\mathbf{R}^3), \quad \text{for} \quad r = 0, 1, 2, \dots,$$

then macroscale quantities are obtained by weighted moments as follows:

$$M_r = M_r[f](t, \mathbf{x}) = \int_{\mathbf{R}^3} m\, \mathbf{v}^r f(t, \mathbf{x}, \mathbf{v})\, d\mathbf{v}, \tag{2.6}$$

where m is the mass of the particles, $r = 0$ corresponds to the local density $\rho(t, \mathbf{x})$ and $r = 1$ to the linear momentum $\mathbf{Q}(t, \mathbf{x})$, while $r = 2$ to the mechanical energy. Therefore, the local density and mean velocity are computed as follows:

$$\rho(t, \mathbf{x}) = m \int_{\mathbf{R}^3} f(t, \mathbf{x}, \mathbf{v})\, d\mathbf{v}, \tag{2.7}$$

$$\mathbf{U} = \mathbf{U}(t, \mathbf{x}) = \frac{1}{\rho(t, \mathbf{x})} \int_{\mathbf{R}^3} m\, \mathbf{v}\, f(t, \mathbf{x}, \mathbf{v})\, d\mathbf{v}, \tag{2.8}$$

while the kinetic translational energy is given by the second-order moment

$$\mathscr{E} = \mathscr{E}(t, \mathbf{x}) = \frac{1}{2\rho(t, \mathbf{x})} \int_{\mathbf{R}^3} m\, (\mathbf{v} - \mathbf{U})^2 f(t, \mathbf{x}, \mathbf{v})\, d\mathbf{v}, \tag{2.9}$$

which, at equilibrium conditions, can be related to the local temperature based on the principle of repartition of energy, which is however valid only in equilibrium conditions.

Suppose now that a mathematical model can be derived to describe the time and space dynamics of f, formally as follows:

$$\mathscr{L} f = \mathscr{N} f, \tag{2.10}$$

where $\mathscr{L}$ and $\mathscr{N}$ are a linear and nonlinear operators, still to be defined explicitly. Then, solving Eq. (2.10) and using Eq. (2.6) yield the desired macroscale information.

Boltzmann's contribution was the derivation of Eq. (2.10) based on some fundamental concepts of non-equilibrium statistical mechanics mixed with some heuristic simplification of physical reality. This celebrated result has generated a huge bibliography not only in the field of physics, but also in that of mathematics due to the challenging hints that this equation induced to applied mathematicians.

The aforementioned mathematical literature tackles a variety of topics including rigorous derivation, qualitative analysis of initial and initial-boundary value problems, and asymptotic limits to derive macroscale equations from the underlying description at the microscopic scale.

However, since dealing with this topic is not the most important aim of this book, we will focus only on those aspects that are relevant to the objective of developing mathematical tools for the derivation of models for living systems. The reader who likes the mathematical kinetic theory also knows that the literature includes some criticism to this model, some of which made difficult the life of the great Boltzmann.

Only the hallmarks followed for the derivation of the Boltzmann equation will be given. As already mentioned, the interested reader is addressed to the book [82] which is, according to the author's bias, the most appropriate for the aforesaid purpose. The reader who is more interested in fluid dynamics applications is addressed to the classical book by Kogan [164], not yet obsolete although appeared more than fifty years ago. The said hallmarks and related technical calculations for the derivation of the model are as follows:

1. Development of detailed calculations of the interaction dynamics of perfectly elastic pair of spherically shaped particles;
2. Computation of the net flow of particles entering into the elementary volume $[\mathbf{x} + d\mathbf{x}] \times [\mathbf{v} + d\mathbf{v}]$ of the phase space and leaving it due to the interaction (collision) dynamics;
3. Modeling the time and space dynamics, in the said elementary volume, for the one-particle distribution function, Eq. (2.5), by equating the local evolution of the distribution function to the net flow of particles due to collisions.

Detailed calculations are reported in the next two subsections.

2.2.1 *Interaction dynamics*

Let us consider binary interactions (collisions) preserving mass, linear momentum, and energy between two particles with equal mass. Moreover, let us denote by $\mathbf{v}$ and $\mathbf{v}_*$ their velocities before interaction (collision) and by $\mathbf{v}'$ and $\mathbf{v}'_*$ those after interaction (collision). The said conservation of mechanical quantities implies:

$$\begin{cases} \mathbf{v} + \mathbf{v}_* = \mathbf{v}' + \mathbf{v}'_*, \\ v^2 + v_*^2 = v'^2 + v'^2_*. \end{cases} \tag{2.11}$$

The system of four scalar equations is not sufficient to compute the six component of the post-collision velocities. Usually, see [82], the post-collision velocities are related to the precollision ones as follows:

$$\begin{cases} \mathbf{v}' = \mathbf{v} + \mathbf{k}(\mathbf{k} \cdot \mathbf{q}), \\ \mathbf{v}'_* = \mathbf{v} - \mathbf{k}(\mathbf{k} \cdot \mathbf{q}). \end{cases} \tag{2.12}$$

where $q = \mathbf{v}_* - \mathbf{v}$ is the relative velocity.
The integration domain of $\mathbf{n}$ is

$$\mathscr{S} = \left\{ \mathbf{k} \in \mathbf{R}^2 : \quad ||k|| = 1, \mathbf{k} \cdot \mathbf{q} \geq 0 \right\}.$$

2.2.2 *The Boltzmann equation*

The Boltzmann equation can be derived by a balance of particles in the elementary volume of the phase space $[\mathbf{x}, \mathbf{x} + d\mathbf{x}] \times [\mathbf{v}, \mathbf{v} + d\mathbf{v}]$. Such a balance can be formally obtained by equating the flow due to transport

$$\mathscr{L} f = (\partial_t + \mathbf{v} \cdot \nabla_{\mathbf{x}} + \mathbf{F}(t, \mathbf{x}) \cdot \nabla_{\mathbf{v}}) f(t, \mathbf{x}, \mathbf{v}), \tag{2.13}$$

to the net flow due to collisions

$$\mathscr{L} f = \mathscr{N} f = G[f, f](t, \mathbf{x}, \mathbf{v}) - L[f, f](t, \mathbf{x}, \mathbf{v}), \tag{2.14}$$

where $\mathbf{F}$ is an external force field, while G and L denote the so-called *gain* and *loss* terms amounting to the inlet and outlet of particles due to collisions.

The calculation of these quantities needs some heuristic assumptions:

1. The probability of collisions involving more than two particles is much smaller than the one of binary encounters;
2. The effect of external forces on the molecules during the mean collision time is negligible in comparison to the interacting molecular forces;
3. Both asymptotic pre-collision and post-collision velocities are not correlated. This hypothesis is referred to as *molecular chaos* and implies that the joint-particle distribution function of the two interacting particles can be factorized, namely is given by the product $f(t, \mathbf{x}, \mathbf{v}) f(t, \mathbf{x}, \mathbf{v}_*)$;
4. The distribution function does not significantly change over a time interval which is larger than the mean collision time but smaller than the mean free time. Likewise, the distribution function does not change very much over a distance of the order of the range of the intermolecular forces.

These assumptions have an impact on the validity of the BE and on the qualitative analysis of the mathematical problems related to its fluid dynamics applications. Many studies have been devoted to these topics and part of them can be regarded as fundamental contributions to applied mathematics. However, these issues are not deeply dealt with in this book, where the BE is simply presented as a conceptual parallel to the mathematical approach hereinafter developed. Additional bibliographic references will be given in the following, in addition to the already cited book [82].

Bearing all above in mind, the terms G and L can be computed explicitly according to the aforementioned assumptions.

Let us consider two gas molecules whose asymptotic pre-collision velocities are denoted by $\mathbf{v}$ and $\mathbf{v}_*$. In Fig. 2.1, the molecule which has velocity $\mathbf{v}$ is at the point O, while the other molecule is approaching with relative velocity $\mathbf{q} = \mathbf{v}_* - \mathbf{v}$ and with impact parameter b in the reference plane forming the azimuthal angle ε with respect to the reference plane.

During a small the time interval Δt all molecules with velocities within the range $\mathbf{v}_*$ and $\mathbf{v}_* + d\mathbf{v}_*$, and that are inside the cylinder of volume $q\Delta t b db d\varepsilon$ will collide with the molecules located in a volume element $d\mathbf{x}$ around the point O and whose velocities are within the range $\mathbf{v}$ and $\mathbf{v} + d\mathbf{v}$. The number of the former is given by $f(t, \mathbf{x}, \mathbf{v}_*)d\mathbf{v}_* q\Delta t b db d\varepsilon$ while the number of the latter is $f(t, \mathbf{x}, \mathbf{v})d\mathbf{x}d\mathbf{v}$. Therefore, the total number of particles which are expected to loose the velocity $\mathbf{v}$ per unit of time in the phase volume element $d\mathbf{x}d\mathbf{v}$ is given by

$$L(f, f)(t, \mathbf{x}, \mathbf{v})d\mathbf{x}\, d\mathbf{v}\Delta t = d\mathbf{x}\, d\mathbf{v}\Delta t \int f(t, \mathbf{x}, \mathbf{v}) f(t, \mathbf{x}, \mathbf{v}_*)\, q\, b\, db\, d\varepsilon\, d\mathbf{v}_*, \quad (2.15)$$

where integration is over the domain of the variables under the integral sign.

However, collisions are reversible and, in the reversed interactions, particles can also assume a velocity between $\mathbf{v}$ and $\mathbf{v} + d\mathbf{v}$ as a result of collisions with the following characteristics:

(i) Asymptotic pre-collision velocities $\mathbf{v}'$ and $\mathbf{v}'_*$;
(ii) Asymptotic post-collision velocities $\mathbf{v}$ and $\mathbf{v}_*$;
(iii) Apsidal vector $\mathbf{k}' = \mathbf{k}$,
(iv) Impact parameter $b' = b$;
(v) Azimuthal angle $\varepsilon' = \pi + \varepsilon$.

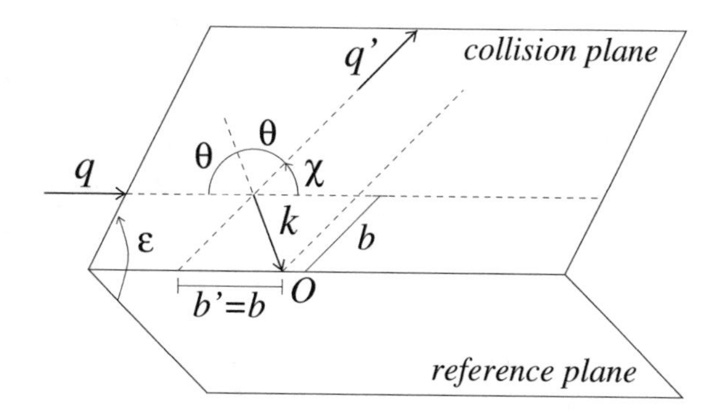

Fig. 2.1 Interaction dynamics. $\mathbf{q}$ and $\mathbf{q}'$ are the pre-collision and post-collision relative velocities, b is the impact parameter, and χ is the scattering angle

These collisions can be referred to as inverse collisions to distinguish them from the former which are called direct collisions. The geometry of direct and inverse collisions is represented in Fig. 2.2.

In analogy with the treatment of the direct collisions, it is not difficult to conclude that the total number of particles which are expected to gain the velocity $\mathbf{v}$ per unit of time in the phase volume element $d\mathbf{x}d\mathbf{v}$ is given by

$$G(f, f)(t, \mathbf{x}, \mathbf{v})d\mathbf{x}\, d\mathbf{v}\Delta t = d\mathbf{x}\, d\mathbf{v}\Delta t \int f(t, \mathbf{x}, \mathbf{v}')f(t, \mathbf{x}, \mathbf{v}'_*)\, q\, b\, db\, d\varepsilon\, d\mathbf{v}_*, \quad (2.16)$$

where it has been used the fact that the modulus of the Jacobian for the equations that relate the post- and pre-collision asymptotic velocities is equal to one, i.e., $d\mathbf{v}'d\mathbf{v}'_* = d\mathbf{v}d\mathbf{v}_*$, and, due to the energy conservation law, the modulus of the pre- and post-collision relative velocities are equal to each other, i.e., $q' = q$.

By using Eqs. (2.16) and (2.15), the Boltzmann equation can be thus written:

$$(\partial_t + \mathbf{v} \cdot \nabla_{\mathbf{x}} + \mathbf{F}(t, \mathbf{x}) \cdot \nabla_{\mathbf{v}})\, f(t, \mathbf{x}, \mathbf{v})$$
$$= \int \left[f(t, \mathbf{x}, \mathbf{v}')f(t, \mathbf{x}, \mathbf{v}'_*) - f(t, \mathbf{x}, \mathbf{v})f(t, \mathbf{x}, \mathbf{v}_*)\right] q\, b\, db\, d\varepsilon\, d\mathbf{v}_*. \quad (2.17)$$

Various modifications of this kinetic equation are known in the literature. Some of them are analyzed in the next section still looking at the development of mathematical tools for living systems.

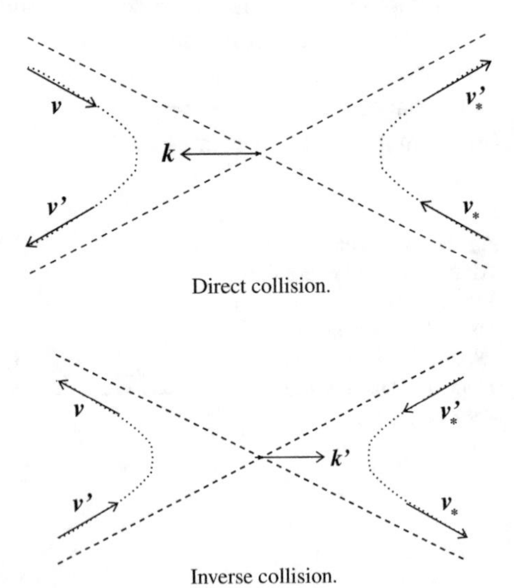

Fig. 2.2 Direct and inverse collisions

2.2.3 Properties of the Boltzmann equation

Although the Boltzmann equation is derived under various approximations of physical reality, it still retains some important features of the physics of system composed of many interacting classical particles. Let us first observe that the local density and mean velocity are computed by Eqs. (2.7) and (2.8), while the local temperature can be related to the kinetic energy by assuming that the principle of repartition of the energy for a perfect gas can be applied.

Bearing this in mind, some important properties can be recalled referring to the classical literature:

1. The collision operator $\mathscr{N}[f, f]$ admits collision invariants corresponding to mass, momentum, and energy, namely

$$< \phi_r, \mathscr{N}[f, f] >= 0, \quad \text{for} \quad r = 0, 1, 2, \quad \text{with} \quad \phi_r = \mathbf{v}^r, \tag{2.18}$$

   where the notation $<,>$ is used for product under the integral sign. Equation (2.18) states conservation of mass, linear momentum, and energy.
2. A unique solution f_e to the equilibrium equation $\mathscr{N}[f] = 0$ exists and is the Maxwellian distribution:

$$f_e(t, \mathbf{x}, \mathbf{v}) = \frac{\rho(t, \mathbf{x})}{[2\pi(k/m)\Theta(t, \mathbf{x})]^{3/2}} \exp\left\{ - \frac{|\mathbf{v} - \mathbf{U}(t, \mathbf{x})|^2}{2(k/m)\Theta(t, \mathbf{x})} \right\}, \tag{2.19}$$

   where k is the Boltzmann constant, and Θ is the temperature given, for a monatomic gas, by the approximation of a perfect gas, namely

$$\mathscr{E} = \frac{3}{2}\frac{k}{m}\Theta,$$

   where $\mathscr{E}$ is the already defined kinetic energy.

The first item states conservation of mass

$$\int f(t, \mathbf{x}, \mathbf{v})\, d\mathbf{x}\, d\mathbf{v} = \int f_0(\mathbf{x}, \mathbf{v})\, d\mathbf{x}\, d\mathbf{v},$$

linear momentum

$$\int \mathbf{v}\, f(t, \mathbf{x}, \mathbf{v})\, d\mathbf{x}\, d\mathbf{v} = \int \mathbf{v}\, f_0(\mathbf{x}, \mathbf{v})\, d\mathbf{x}\, d\mathbf{v},$$

and energy

$$\int v^2\, f(t, \mathbf{x}, \mathbf{v})\, d\mathbf{x}\, d\mathbf{v} = \int v^2\, f_0(\mathbf{x}, \mathbf{v})\, d\mathbf{x}\, d\mathbf{v},$$

where $f_0(\mathbf{x}, \mathbf{v}) = f(t = 0, \mathbf{x}, \mathbf{v})$.

Trend to equilibrium is assured by the *kinetic entropy*, namely the H-Boltzmann functional:

$$H[f](t) = \int_{\mathbf{R}^3 \times \mathbf{R}^3} (f \, log f)(t, \mathbf{x}, \mathbf{v}) \, d\mathbf{x} \, d\mathbf{v}, \tag{2.20}$$

which, in the spatially homogeneous case, is monotonically decreasing along the solutions and is equal to zero at $f = f_e$. Therefore, the Maxwellian distribution with parameters ρ, $\mathbf{U}$, and Θ minimizes H (this is the celebrated H-Theorem).

Mathematical problems for the Boltzmann equation can be classified as the initial value (Cauchy) problem in unbounded domains and the initial-boundary value problem in bounded domains or external flows. Applications refer generally to non-equilibrium thermodynamics or fluid dynamics applications for molecular flows. The *Cauchy problem* can be stated in more detail, in the absence of an external force, as follows:

$$\begin{cases} \partial_t f + \mathbf{v} \cdot \nabla_{\mathbf{x}} f = J(f, f), \\ \\ f(t = 0, \mathbf{x}, \mathbf{v}) = f_0(\mathbf{x}, \mathbf{v}), \end{cases} \tag{2.21}$$

where the initial datum can either be assumed to decay at infinity in the phase space or be periodic in space and decaying with velocity.

The literature on this challenging problem is documented in various books, for instance [13, 49, 131], which report about the solution of Cauchy problem for initial data given as perturbation of vacuum, spatially homogeneous case, and Maxwellian equilibrium. In particular, the book by Glassey [131] reports also about the Cauchy problem for the mean field models, namely the Vlasov equation.

Particularly important is the celebrated global existence result by Di Perna and Lions [99, 100], who succeeded to prove existence in the sense of distributions for a large class of initial data for a gas with bounded mass momentum and energy. This result gave, as it is known, a great hint to research activity as all previous results were limited to small perturbation although the existence theorem [53] also included the case of an infinite mass of the gas.

The statement of the initial-boundary value problem requires the modeling of gas–surface interaction dynamics. Two specific problems, among various ones, can be stated:

- The *interior domain problem*, which corresponds to a gas contained in a volume bounded by a solid surface;
- The *exterior domain problem*, which corresponds to a gas in the whole space $\mathbf{R}^3$ which contains an obstacle.

The surface of the solid wall is defined in both cases by Σ_w, and the normal to the surface directed toward the gas is ν. Moreover, in order to define the boundary conditions on a solid wall, we need to define the partial incoming and outgoing traces f^+ and f^- on the boundary Σ_w, which, for continuous f, can be defined as follows:

$$f^+(\mathbf{x}, \mathbf{v}) = f(\mathbf{x} \in \Sigma_w, \mathbf{v} | \mathbf{v} \cdot \nu(\mathbf{x}) > 0), \tag{2.22}$$

and

$$f^-(\mathbf{x}, \mathbf{v}) = f(\mathbf{x} \in \Sigma_w, \mathbf{v}|\, \mathbf{v} \cdot \nu(\mathbf{x}) < 0). \tag{2.23}$$

The boundary conditions can be formally defined by the operator map:

$$f^+(t, \mathbf{x}, \mathbf{v}) = \mathscr{R}\, f^-(t, \mathbf{x}, \mathbf{v})\,, \tag{2.24}$$

where the operator $\mathscr{R}$, which maps the distribution function of the particles which arrive to the surface into the one of particles leaving the surface, is featured by the following properties:

1. $\mathscr{R}$ is linear, of local type with respect to **x**, and is positive:

$$f^- \geq 0 \Rightarrow \mathscr{R}\, f^- \geq 0. \tag{2.25}$$

2. $\mathscr{R}$ preserves mass, i.e., the flux of the incoming particles equals the one of the particles which leaves the surface.
3. $\mathscr{R}$ preserves local equilibrium at the boundary: $\omega_w^+ = \mathscr{R}\, \omega_w^-$, where ω_w is the Maxwellian distribution at the wall temperature.
4. $\mathscr{R}$ is dissipative, i.e., satisfies the inequality at the wall:

$$\int_{\mathbf{R}^3} \langle \mathbf{v} \cdot \nu \rangle \left(f^- + \mathscr{R}\, f^-\right) \left(\log f + |\mathbf{v}|^2/T\right)\, d\mathbf{v} \leq 0. \tag{2.26}$$

The formulation of the initial-boundary value problem, in the case of the *interior domain problem*, consists in linking the BE to the initial and boundary conditions. In the case of the *exterior domain problems*, in addition to the boundary conditions on the wall, suitable Maxwellian equilibrium conditions need to be assumed at infinity. The already cited book [82] reports some specific examples of the operator $\mathscr{R}$.

2.3 Some Generalized Models

The mathematical developments of the kinetic theory have generated a variety of models somehow related to the original Boltzmann equation. Most of them are motivated by the need of reducing the analytic and computational complexity of the equation, while others aim at taking into account physical effects that are not included in the original model. This section focuses on three different classes of models, selected among various ones, which present features of interest for a comparison with the approach addressed to living systems that will be presented in the next chapter.

Specifically, we consider the so-called BGK model, where the complexity of the right-hand-side term modeling collisions is strongly simplified by replacing the original term by an algebraic structure modeling, a trend toward equilibrium. The second class of models corresponds to the so-called Discrete Boltzmann equation, where particles are supposed to attain only a finite number of velocities. Finally, the Enskog equation is rapidly introduced as a model, which attempts to include some dense gas effects. The next three paragraphs present these models waiting for the critical analysis proposed in the last section.

2.3.1 *The BGK model*

This model was proposed, as already mentioned, to reduce the analytic and computational complexity of the Boltzmann equation. The hallmarks that guided the design of this model can be listed as follows:

- If the distribution function f is known, then the local Maxwellian can be computed by the first three moments. Formally: $f_e = f_e[f] =: f_e(\rho, \mathbf{U}, \Theta)$.
- A molecular fluid is featured by a natural trend to local equilibrium as described by the H-function.
- Models can be derived simply by replacing the collision right-hand-side integral operator by a decay term modeling trend to equilibrium.

Consequently, in the absence of an external force field, the model is written as follows:

$$\partial_t f + \mathbf{v} \cdot \nabla_{\mathbf{x}} f = c[f](f_e[f] - f), \tag{2.27}$$

where c gives the decay rate. The latter can be assumed to be a function of the local density ρ, i.e., $c = c(\rho)$. Otherwise, the simplest approximation suggests to take a constant value $c = c_0$.

Model (2.27) definitely requires less sophisticated tools toward computing than the original Boltzmann equation. However, one has necessarily to put in discussion the validity of the decay assumptions in conditions largely far from equilibrium.

In addition, let us stress that the BGK model is nonlinear even in the case of $c = c_0$. In fact, f_e nonlinearly depends on f. It becomes a simple linear model only by linearization or by assuming that $f_e = f_{e0}$, which is rather unrealistic hypothesis.

The substantial conceptual difference with respect to the Boltzmann equation is that in the original model, the Maxwellian distribution is an emerging behavior that is induced by the dynamics at the microscopic scale, while it is artificially imposed in the BGK model.

2.3.2 *The discrete Boltzmann equation*

Discrete velocity models of the Boltzmann equation can be obtained, assuming that particles allowed to move with a finite number of velocities. The model is an evolution equation for the number densities N_i linked to the admissible velocities $\mathbf{v}_i$, for

$i = 1, \ldots, n$ with $i \in L = \{1, \ldots, n\}$. The set $\{N_i(t, \mathbf{x})\}_{i=1}^n$ corresponds, for certain aspects, to the one-particle distribution function of the continuous Boltzmann equation.

The mathematical theory of the discrete kinetic theory is exhaustively treated in the Lecture Notes by Gatignol [125], which provides a detailed analysis of the relevant aspects of the discrete kinetic theory: modeling, analysis of thermodynamic equilibrium, and application to fluid-dynamic problems. The contents mainly refer to a simple monatomic gas and to the related thermodynamic aspects.

After such a fundamental contribution, several developments have been proposed in order to deal with more general physical systems: gas mixtures, chemically reacting gases, particles undergoing multiple collisions, and so on. Analytic topics, such as the qualitative analysis of the initial value and of the initial-boundary value problem, have been object of continuous interest of applied mathematicians as reported in the survey [45]. Additional sources of information are the review paper by Platkowski and Illner [201] and the edited book [42], where various applications and developments of the model are reported.

The formal expression of the evolution equation corresponds, as for the full Boltzmann equation, to the balance of particles in the elementary volume of the space of the microscopic states:

$$(\partial_t + \mathbf{v}_i \cdot \nabla_{\mathbf{x}})\, N_i = J_i[N] = \frac{1}{2} \sum_{j,h,k=1}^{n} A_{ij}^{hk} (N_h N_k - N_i N_j), \tag{2.28}$$

which is a system of partial differential equations, where the dependent variables are the number of densities linked to the discrete velocities:

$$N_i = N_i(t, \mathbf{x}) \; : \quad (t, \mathbf{x}) \in [0, T] \times \mathbf{R}^{\nu} \to \mathbf{R}_+ , \quad i = 1, \ldots, n , \quad \nu = 1, 2, 3.$$

Collisions $(\mathbf{v}_i, \mathbf{v}_j) \longleftrightarrow (\mathbf{v}_h, \mathbf{v}_k)$ are binary and reversible, and preserve mass momentum and energy. Their modeling is left to the so-called *transition rates* A_{ij}^{hk}, which are positive constants and, according to the reversibility properties, satisfy the following relations: $A_{ji}^{hk} = A_{ij}^{kh} = A_{ji}^{kh}$.

Analogously to the Boltzmann equation, it is possible to define the space of collision invariants and of the Maxwellian state [125]. In more detail, the following definitions can be used:

Collision Invariant: A vector $\phi = \{\phi_i\}_{i\in L} \in \mathbf{R}^n$ is defined *collision invariant* if:

$$\langle \phi \, , \, J[N] \rangle = 0 \, , \quad J[N] = \{J_{i\in L} \in \mathbf{R}^n\},$$

where the inner product is defined in $\mathbf{R}^n$.

Space of the collision invariants: The set of the totality of collision invariants is called *space of the collision invariants* and is a linear subspace of $\mathbf{R}^n$. Such a space will be denoted, in what follows, by $\mathscr{M}$.

Maxwellian: Let $N_i > 0$ for any $i \in L$, then the vector N is defined *Maxwellian* if $J[N] = 0$. In particular, let $N \in \mathbf{R}^n$, then the following conditions are equivalent:
(i) N is a Maxwellian;
(ii) $\{\log N_i\}_{i \in L} \in \mathcal{M}$;
(iii) $J[N] = 0$.

Applied mathematicians have attempted in the last decade to design models with arbitrarily large number of velocities and hence to analyze convergence of discretized models toward the full Boltzmann equation. However, various technical difficulties have to be tackled, such as:

1. Discretization schemes for each couple of incoming velocities do not assure a pair of outgoing velocities such that conservation of mass and momentum is preserved;
2. Specific models may have a number of spurious collision invariants in addition to the classical ones corresponding to conservation of mass, linear momentum, and energy;
3. Convergence of the solutions of discretized equation to those of the full Boltzmann equation, when the number of discretization points tends to infinity, under suitable hypotheses which have to be properly defined.

A technical difficulty in dealing with the above convergence proof consists in obtaining an existence theorem for the Boltzmann equation in a function space which can be properly exploited for the development of the computational scheme.

These difficulties generated the so-called $semi-continuous$ $Boltzmann$ $equation$, where particles are allowed to attain all directions in space, but only a finite number of velocity modules. This approach allows to overcome the difficulty in integration of the whole velocity space, namely $\mathbf{R}^3$. Particularly important is the result of paper [171], where it is shown that a particular choice of the sequence of velocity modules maximizes the number of collisions consistent with conservation of momentum and energy; hence, the model appears closer to physical reality.

2.3.3 Vlasov and Enskog equations

Further models have been developed to take into account some alternative ways of modeling interactions. For instance, the Boltzmann equation is such that the distribution function is modified only by external actions and short range interactions and, on the other hand, various physical systems are such that also long-range interactions may be significant as in the mean field approaches which lead to the Vlasov equation. Moreover, one can consider localized interactions, but at a fixed distance imposed by the finite dimension of the particles.

Both types of interactions can occur in the case of living systems although in a technically different way. Therefore, it is worth providing a brief introduction to the said alternative derivation methods.

Focusing on the mean field approach, let us consider the vector action $\mathcal{P} = \mathcal{P}(\mathbf{x}, \mathbf{v}, \mathbf{x}_*, \mathbf{v}_*)$ on the subject with microscopic state $\mathbf{x}, \mathbf{v}$ (test particle) due to the subject with microscopic state $\mathbf{x}_*, \mathbf{v}_*$ (field particle). The resultant action is:

$$\mathcal{F}[f](t, \mathbf{x}, \mathbf{v}) = \int_{\mathbf{R}^3 \times \mathcal{D}_\Omega} \mathcal{P}(\mathbf{x}, \mathbf{v}, \mathbf{x}_*, \mathbf{v}_*)\, f(t, \mathbf{x}_*, \mathbf{v}_*)\, d\mathbf{x}_*\, d\mathbf{v}_*, \tag{2.29}$$

where $\mathcal{D}_\Omega$ is the domain around the test particle, where the action of the field particle is effectively felt; namely, the action $\mathcal{P}$ decays with the distance between test and field particles and is equal to zero on the boundary of $\mathcal{D}_\Omega$.

Based on the aforesaid assumptions, the mean field equation writes:

$$\partial_t f + (\mathbf{v} \cdot \nabla_\mathbf{x}) f + \mathbf{F} \cdot \nabla_\mathbf{v} f + \nabla_\mathbf{v} \cdot (\mathcal{F}[f]\, f) = 0, \tag{2.30}$$

where $\mathbf{F}$ is the positional macroscopic force acting on the system.

Another model to be considered is the so-called *Enskog equation*, which introduces some effects of the finite dimensions of the particles. This specific feature is introduced in the model by two ways:

1. The distribution functions of the interacting pairs (indeed, only binary collisions are considered) are computed in the centers of the two spheres, namely not in a common point for both distributions;
2. The collision frequency is reduced by the dimension of the interacting spheres, which reduces the probability of further interactions by shielding the free volume available for further interactions. This amounts to introducing a functional local density which correlates with the distribution functions of the interacting pairs and increases with the increasing local density.

This model has been interpreted for a long time as the first step toward the derivation of kinetic models for dense fluid. On the other hand, the limitation to binary mixtures technically prevents such interesting generalization. In fact, multiple interactions appear to be necessary to describe the physics of transition from rarefied to dense fluid.

However, as it is, the model offers to applied mathematicians a a variety of challenging problems such as existence and uniqueness of solutions to mathematical problems and asymptotic limits either to hydrodynamics when the intermolecular distances tends to zero, or to the Boltzmann equation when the dimension of the particles is allowed to tend to zero. The book [48] was devoted to this model and mainly to the analytic problems generated by its applications. The bibliography reported in it also includes useful indications in the field of physics.

2.4 Computational Methods

Kinetic equations have the form of nonlinear integro-differential equations in which the unknown function depends on seven variables. The mathematical difficulties connected with them are such that only numerical solutions can be obtained in cases of practical interest. The common strategy consists in decoupling the transport and the collision terms by time-splitting the evolution operator into a *drift step*, in which collisions are neglected, and a *collision step*, in which spatial motion is frozen.

Numerical methods can be roughly divided into three groups depending on how drift and collision steps are dealt with, namely regular, semi-regular, and particle. Valuable references are provided by standard books [10, 64, 197, 208] as well as recent review papers [106].

Regular and semi-regular methods adopt similar strategies in discretizing the distribution function on a grid in the phase space [10]. The drift step requires to solve a system of hyperbolic conservation laws coupled at the boundaries. Their discretization can be performed in a variety of ways, including finite-difference, finite-volume, finite-element, or spectral methods [170]. The collision step consists of solving a spatially homogeneous relaxation equation. This is the more computationally demanding part since it involves the computation of a high-dimensional integral defining the collision operator. Regular and semi-regular methods differ in the way the collision term is evaluated.

Most of the regular methods adopt a Galerkin discretization of the velocity space [106]. These methods consist of expanding the velocity dependence of the distribution function in a set of trial functions with expansion coefficients that depend on position and time. The Galerkin ansatz is substituted in the space homogeneous relaxation equation which is subsequently multiplied by test functions and integrated into the velocity space. According to the Galerkin approach, test and trial functions are assumed to be the same. The above procedure yields a coupled system of ordinary differential equations for the expansion coefficients. Galerkin discretization can be further distinguished depending on the basis functions which they employ.

In Fourier–Galerkin approach, the distribution function is expanded in trigonometric polynomials and the fast Fourier transform is used to accelerate the computation of the collision integral in the velocity space [114, 122, 178, 196], while discontinuous Galerkin methods adopt discontinuous piecewise polynomials as test and trial functions [9, 126].

Hybrid approaches have been also developed where the distribution function is expanded in Laguerre polynomials with respect to the velocity components parallel to solid surfaces, whereas quadratic finite-element functions have been used for the normal velocity component [190, 225].

In semi-regular methods, the collision integral is computed by Monte Carlo or quasi-Monte Carlo quadrature. These schemes originate from the work by Nordsieck and Hicks [185] and have been further developed over the years by a number of authors [11, 116].

Particle methods originate from the Direct Simulation Monte Carlo (DSMC) scheme [64] which, introduced firstly based on physical reasoning [65], has been later proved to converge, in a suitable limit, to the solution of the Boltzmann equation [238]. The distribution function is represented by a number of particles which move in the computational domain and collide according to stochastic rules derived from the kinetic equations.

The space domain to be simulated is covered by a mesh of cells. These cells are used to collect together particles that may collide and also for the sampling of macroscopic field such as density and mean velocity. Macroscopic flow properties are obtained by time averaging particle properties. Variants of the DSMC have been proposed over the years which differ in the way the collision step is performed.

These methods include the majoring Frequency scheme [158], the null Collision scheme [165], the Nanbu scheme [182] and its modified version [17], the Ballot Box scheme [249], the Simplified Bernoulli trials scheme [227].

Particle schemes and particle methods are by far the most popular and widely used simulation methods in rarefied gas dynamics. Indeed, they permit to easily handle complex geometries while keeping the computing effort requirements at a reasonable level. Furthermore, they permit to easily insert new physics.

Particle methods are not well suited to simulate unsteady gas flows. Indeed, in this case, the possibility of time averaging is lost or reduced. Acceptable accuracy can only be achieved by increasing the number of simulation particles or superposing several flow snapshots obtained from statistically independent simulations of the same flow but, in both cases, the computing effort is considerably increased. Weighted particles can also be used to increase the result accuracy without additional computational cost [208].

Although the adoption of such a grid limits the applicability of regular and semi-regular schemes to problems where particular symmetries reduce the number of spatial and velocity variables, these are a feasible alternative to particle methods in studying unsteady flows since the possibility of a direct steady-state formulation.

Moreover, unlike particle-based methods, their implementation on massively parallel computers with Single Instruction Multiple Data (SIMD) architecture, such as multicores and Graphic Processing Units (GPUs), can easily realize the full potential of these processors [119, 120]. These aspects also prompted the development of deterministic methods of solution.

2.5 Critical Analysis

Let us now elaborate a critical overview of the contents of our book referring it to the complexity features presented in Chapter 1. Criticisms should not be addressed to the mathematical kinetic theory as it is. Indeed, it remains a fundamental research field of applied mathematics in interaction with physics and, as far as some applications are concerned, with technological sciences. On the other hand, we wish to understand the drawbacks of kinetic theory methods when referred to the modeling of living systems.

The aim is also to avoid an uncritical use of classical methods and to understand how the classical methods should be modified to tackle the challenging task of capturing the aforesaid complexity features.

Bearing all above in mind, it can be rapidly understood that the mathematical structures of the kinetic theory do not retain the main features of complex systems, where the causality principles which rule the dynamics of classical particles are lost. On the other hand, it can be argued that the representation by a probability distribution over the microscopic state can provide a useful description of the whole system. In fact, it retains some stochastic features of living systems and can rapidly provide, simply by weighted moments, the information on macroscale properties, which are often needed by applications.

An example can be rapidly selected to understand the conceptual differences between classical and living particles. Then, let us consider the statement of the boundary conditions as in Eq. (2.25). Classical particles collide with the wall where they sharply modify their trajectory. Living particles feel (see) the wall and gradually modify their trajectory before reaching it.

Furthermore, the microscopic state cannot be limited to mechanical variables as it should capture, in an appropriate way, the ability that interacting entities have to express their strategy.

In addition, the use of a distribution function can be made technically consistent with the presence of a birth and death dynamics, which is a typical feature of living systems including features not precisely related to classical mechanics. For instance, birth processes can even generate mutations and selection within a Darwinian-type framework [57].

According to all above reasonings, it can be argued that the idea of representing the system by some modified (generalized) distribution function can be accepted as a useful step toward modeling. Moreover, the balance of particles in the elementary volume of the space of the microscopic states can still be used to compute the time and space dynamics of the aforesaid distribution function. However, interactions cannot be described by the deterministic causality principles of classical mechanics, starting from the fact that interactions are not reversible. Therefore, new ideas need to be developed.

The rationale that guides the approach of this book is that theoretical tools of game theory need to be properly developed to model interactions. Of course, one cannot naively follow a straightforward application of the classical game theory, but has to look for developments suitable to include multiple interactions and learning ability, which has a progressive influence on the rules of interactions that evolve in time [186, 187]. Furthermore, modeling interactions should necessarily taken into account stochastic aspects, which are somehow related to the heterogeneous behavior of living entities, and space nonlocality, as when interactions can occur in networks [162, 163, 184].

Chapter 3
On the Search for a Structure: Toward a Mathematical Theory to Model Living Systems

3.1 Plan of the Chapter

The contents of this chapter are motivated by the second key question posed in Section 1.3. This question focuses on the search of mathematical tools suitable to model living systems:

Can appropriate mathematical structures be derived to capture the main features of living systems?

The strategy to pursue this challenging objective consists in designing a general mathematical framework (structure) suitable to capture, within a differential system, the ten key complexity features presented in Section 1.4. This general framework offers the conceptual basis for the derivation of specific models and substitutes the field theories that cannot be applied in the case of living systems.

Mathematical models are derived by implementing the said equations with a detailed interpretation, and modeling, of interactions at the microscopic scale, namely at the individual-based level. The rules that guide these interactions can differ from system to system, while the aforesaid mathematical structure transfers the dynamics at the scale of individuals into that of collective behaviors.

A key feature of the modeling approach is the subdivision into groups, later defined functional subsystems. The modeler should develop, by taking advantage of her/his experience and bias, the subdivision which is more effective toward the mathematical description of the system, depending also on the specific analysis under consideration.

This methodological approach can be also applied to systems in a network of nodes, where the distribution within each node is homogeneous in space, while the role of the space variable simply consists in assessing the localization of the nodes, but it is not a continuous variable. Hence, we will use the term *vanishing space variable* to denote this type of dynamics. On the other hand, when space can be treated as a

N. Bellomo et al., *A Quest Towards a Mathematical Theory of Living Systems*, Modeling and Simulation in Science, Engineering and Technology,
DOI 10.1007/978-3-319-57436-3_3

continuous variable, then interactions, and hence the mathematical structure, have to account for this specific feature.

Bearing all above in mind, the approach leads to a general mathematical theory that can generate specific models in different fields of life and social sciences. The modeling of individual-based interactions is developed by theoretical tools of game theory that are typical of the science field, where each model refers to. The overall approach is known as the *kinetic theory of active particles*, for short KTAP theory, where the interacting entities are called *active particles* due to their ability to express a certain strategy which is called *activity*.

These reasonings address our mind to the methods of statistical mechanics and, in particular, of the kinetic theory, although substantial conceptual differences will rapidly appear. Let us now specifically compare the approach to the methods of the kinetic theory of diluted gases and, specifically, to the celebrated Boltzmann equation reviewed in Chapter 2, where it has been mentioned that a common feature shared by the classical and the new approach is that the state of the overall system is described by a probability distribution over the microscopic scale state of the interacting entities. However, interactions are not reversible and involve multiple entities in the case of living systems.

A further aspect which distinguishes the classical and the new approach consists in the rationale followed in the modeling of interactions. In fact, the theory of Newtonian mechanics, namely conservation of mass, momentum, and energy, is applied in the classical theory, while more sophisticated approaches might include quantum and relativistic effects.

Different approaches have to be taken into account in the modeling of interactions among living entities. In fact, these are driven by individual rational behaviors which aim at pursuing the individual well-being of the interacting entities. However, this search is occasionally contrasted by irrational behaviors which might occasionally lead to negative effects although generated by a different wish.

After these preliminaries, details of the content of this chapter can now be given:

- Section 3.2 focuses on large systems, where space is a vanishing variable and it shows how the representation can be given by probability distributions over the microscopic state of the interacting entities. The approach includes also the case of networks, where the distribution within each node is, however, homogeneous in space.

- Section 3.3 shows how interactions can be modeled by theoretical tools of game theory for systems, where the dynamics is homogeneous in space. Then, these models are used to derive the aforesaid general mathematical structure suitable to describe, by a differential system, the time dynamics of the probability distribution which represents the overall state of the system.

- Section 3.4 shows how the approach presented in Section 3.3 can be technically generalized to a network of interacting nodes. In both cases, the dependent variables are the probability distributions over the microscopic states of the interacting entities called active particles.

- Section 3.5 develops an analogous study for models where space is a continuous variable. Interactions take into account specific features related to space continuity, namely the definition of the visibility and sensitivity zones and their influence on the general structure, which is derived in this section.

- Section 3.6 firstly analyzes the consistency of the aforesaid mathematical structures, derived in the preceding sections, with the complexity features reported in Chapter 1. Subsequently, some perspective ideas toward the derivation of a mathematical theory for complex systems are anticipated in view of a deeper analysis proposed in Chapter 7.

3.2 A Representation of Large Living Systems

This section deals with the representation of a large system of living entities, viewed as *active particles*. Such system is complex according to the paradigms reported in Chapter 1. The said representation defines the dependent variables deemed to describe the dynamics of the said system. As already mentioned, these particles have the ability to express a strategy called *activity*, which can heterogeneously differs from particle to particle. Particles that express, however heterogeneously, the same activity are grouped into the so-called *functional subsystems*.

In general, living entities can express a large variety of activities. However, due to the need of reducing the technical complexity induced by an excessive number of equations, we restrict the modeling approach to those specific activities that are object of study. Moreover, we consider only cases, where the activity within each functional subsystem is a scalar. This somehow restrictive assumption is critically analyzed and relaxed, in the next chapters referring to some specific examples.

Let us consider a system of *interconnected networks*, where the number of nodes is constant in time and all of them can interact. In addition, the dynamics within each node is homogeneous in space. Therefore, the space variable simply acts for the localization of the nodes. The representation is proposed as follows:

- The overall system is constituted by n *interconnected nodes* labeled by the subscript $i = 1, \dots, n$.
- In each node, particles are grouped into m *functional subsystems* labeled by the subscript $j = 1, \dots, m$.
- Space is treated as a *vanishing variable*; namely, it is confined to the localization of nodes, while the dynamics in each node is homogeneous in space.
- The activity u is a scalar with values in a domain D_u supposed to be the same for all functional subsystems. Typical values, depending on each system, are $[0, 1]$, $[-1, 1]$, $I\!R_+$, and $I\!R$. Generally, the ability to express a certain activity increases with increasing values of u, while negative values correspond to an expression opposite to the said activity.
- The activity is heterogeneously distributed over D_u, while the overall state of the system is delivered by the probability distribution function

$$f_{ij} = f_{ij}(t,u): \qquad [0,T] \times D_u \to I\!R_+, \tag{3.1}$$

for each ij-functional subsystem and where, under local integrability conditions, $f_{ij}(t,u)du$ denotes the number of particles that at time t have an activity in the elementary domain $[u, u+du]$ of the space of the microscopic states of the ij-functional subsystem.

- Each subsystem can be subject to external actions that can act both at the microscopic and at the macroscopic scales.

External actions at microscopic scale can be viewed as applied by *agents* with known distribution functions $g_{ij} = g_{ij}(t,w)$, with w acting on the same domain of the activity u. It can be useful, in applications, to assume that the external action is given by known product contributions over time and activity, such as

$$g_{ij}(t,w) = \psi_{ij}(t)\,\phi_{ij}(w), \tag{3.2}$$

where integrability conditions need to be applied to ϕ.

External actions can be applied also at a macroscopic level, namely on each functional subsystem as a whole. In this case, the following notation is used $K_{ij} = K_{ij}(t,u)$. Both actions are supposed to be known and not modified by interactions with active particles. Occasionally, the system of active particles is called *inner (internal) system*, while the agents acting on particles are viewed as an *outer (external) system*.

These hallmarks basically refer to [163] and to Figure 3.1, which represents the network of the overall system, where nodes are represented by solid circles and are connected by edges and squiggly arrows represent the external agents.

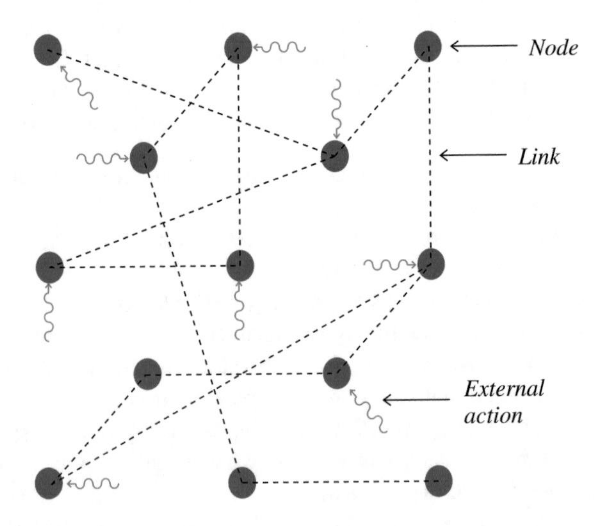

Fig. 3.1 Network of functional subsystems

The representation is transferred to the macroscopic scale by standard moment calculations:

1. If $f_{ij} \in L_1(D_u)$, the number of particles that at time t belong to each ij-functional subsystem is given by:

$$n_{ij} = n[f_{ij}](t) = \int_{D_u} f_{ij}(t, u)\, du, \tag{3.3}$$

while the total number of particles is

$$N[\mathbf{f}](t) = \sum_{i=1}^{n} \sum_{j=1}^{m} n[f_{ij}](t), \tag{3.4}$$

where $\mathbf{f}$ is the set of all f_{ij}, i.e., $\mathbf{f} = \{f_{ij}\}$.

2. If $u^q f_{ij} \in L_1(D_u)$ for $q = 1, 2, \dots$ then moments at time t are computed as follows:

$$\mathbb{E}^q[f_{ij}](t) = \frac{1}{n[f_{ij}](t)} \int_{D_u} u^q\, f_{ij}(t, u)\, du. \tag{3.5}$$

3. If $q = 1$ and $u f_{ij} \in L_1(D_u)$, then first-order moments identify the *activation*

$$A[f_{ij}](t) = \int_{D_u} u\, f_{ij}(t, u)\, du, \tag{3.6}$$

and *activation density*:

$$\mathscr{A}[f_{ij}](t) = \frac{1}{n[f_{ij}](t)} \int_{D_u} u\, f_{ij}(t, u)\, du. \tag{3.7}$$

4. Second-order moments represent the *activation energy* and the *energy activation density*, respectively

$$E[f_{ij}](t) = \int_{D_u} u^2\, f_{ij}(t, u)\, du, \tag{3.8}$$

and

$$\mathscr{E}[f_{ij}](t) = \frac{1}{n[f_{ij}](t)} \int_{D_u} u^2\, f_{ij}(t, u)\, du. \tag{3.9}$$

5. Third-order moments estimate the *distortion*, namely the deviation of the probability distribution from the symmetry of a Gaussian distribution.
6. Moments of the external actions can be computed by calculations analogous to those we have seen above, in particular:

$$n^e_{ij} = n^e[g_{ij}](t) = \int_{D_w} g_{ij}(t, w)\, dw, \tag{3.10}$$

and

$$E^e[g_{ij}](t) = \frac{1}{n^e[g_{ij}](t)} \int_{D_w} w\, g_{ij}(t, w)\, dw, \tag{3.11}$$

which represent, respectively, the number density and the activation density of agents.

Remark 3.1 *If n_{ij} is constant in time, namely $n_{ij}(t) = n_{ij}^0$, then each f_{ij} can be divided by n_{ij}^0 and regarded as a probability density. Moreover, if $N = N_0$ is constant, then the sum of all f_{ij} can be divided by N_0 and can also be regarded as a probability density.*

Remark 3.2 *A simple case is when only one functional subsystem appears in each node. In this case, the overall state is simply delivered by f_i and moment calculations yield:*

$$n[f_i](t) = \int_{D_u} f_i(t, u)\, du, \tag{3.12}$$

and

$$\mathbb{E}^q[f_i](t) = \frac{1}{n[f_i](t)} \int_{D_u} u^q\, f_i(t, u)\, du. \tag{3.13}$$

Remark 3.3 *It can be useful dealing with discrete activity variables, say $u \in I_u = \{u_1, \ldots, u_r, \ldots, u_p\}$. So that the representation is a discrete probability distribution that for each functional subsystem writes*

$$f_{ij}^r = f_{ij}(t; u_r): \qquad [0, T] \to I\!R_+, \tag{3.14}$$

where $r = 1, \ldots, p$. Moments are computed by the same calculations presented above, where integrals are substituted by finite sums. For instance, moments are computed as follows:

$$\mathbb{E}^q[\mathbf{f}](t) = \frac{1}{n_{ij}[\mathbf{f}](t)} \sum_{r=1}^{p} u_r^q\, f_{ij}^r(t), \tag{3.15}$$

where

$$n_{ij}[\mathbf{f}](t) = \sum_{r=1}^{p} f_{ij}^r(t). \tag{3.16}$$

It is worth stressing that the use of discrete activity variables is not motivated by reducing computational complexity of the equation with continuous states, which was one of the motivations of the discrete Boltzmann equation [42]. It needs to be regarded, in the mathematical approach of the kinetic theory for active particles, as a method to replace the assumption of continuity of the probability distribution over the microscopic state for systems where the number of particles is not large enough to justify it. Some of the applications treated in the next chapters will discuss and critically analyze the choice of discrete variables.

Some additional considerations, to be made more precise looking ahead to specific applications, are the following:

Remark 3.4 *The splitting into functional subsystems differs from classical approaches where the overall systems are featured by physical components well localized in space. Here, each subsystem is featured by the specific functions expressed in the specific dynamics under consideration.*

Remark 3.5 *The splitting into functional subsystems, within the aforesaid system approach, has not a universal meaning. In fact, it depends upon the specific investigation under consideration. Therefore, different modeling perspectives correspond to different strategies to decompose the system. In some cases, the use of a scalar activity variable simplifies the mathematical structure of the model. However, it cannot always be put in practice, and vector activity variables might be needed.*

3.3 Mathematical Structures for Systems with Space Homogeneity

This section presents the derivation of mathematical structures suitable to capture the complexity features of large systems interacting in a node by a dynamics homogeneous in space. The presentation is in five subsections. In more detail, the first three subsections deal with systems in absence of external actions: Firstly a brief phenomenological description is proposed for the games that can be used to model interactions; then, the operators that can describe the modeling are defined; finally, the mathematical framework is derived. Two additional topics complete the presentation, namely the dynamics in the presence of external actions and the analysis of the sources of nonlinearity. The contents of this section also support the next one devoted to the dynamics over a network and derivation of structures for systems with discrete activity variables.

3.3.1 A phenomenological description of games

Interactions are modeled by theoretical tools of game theory. Living entities, at each interaction, *play a game* with an output that technically depends on the strategy they express. This strategy is often related to surviving and adaptation abilities, namely to an individual or collective search for well-being [52]. The output of the game is not deterministic, while the rules of interactions evolve in time due to various reasons such as learning ability, mutating conditions of the environment, which can even be modified by the interaction with particles. This dynamics is also related to the feedback that active particles receive from their environment, which modifies the strategy they express adapting it to mutating environmental conditions [173, 174].

Let us introduce some ideas borrowed from game theory [25, 76, 130, 186, 189, 214, 215, 218] and select certain typologies of games. Their modeling provides information on the dynamics at the microscopic scale that can be introduced, as we shall see in the next subsection, into a general mathematical structure suitable to provide the overall dynamics. The description is here proposed simply at a qualitative level leaving to the next subsections the analytic formalization. In detail, let us consider the following types of games:

1. *Competitive (dissent) games* : The interacting particle with higher state increases its status by taking advantage of the other, which is obliged to decrease it. Therefore, the competition brings advantage to only one of the two players, namely one gains, while the other looses. This type of interaction has the effect of increasing the difference between the states of interacting particles, due to a kind of driving back effect. In most cases, the amount of exchanged microscopic state should remain constant. Namely, what is gained by one side is lost by the other side. However, in a more general framework, some specific interactions can be dissipative or productive.
2. *Cooperative (consensus) games* : Interacting particles show a trend to share their microscopic state; namely, the particle with higher state decreases its status with advantage of the other, which increases it. Therefore, such type of interaction leads to a decrease in the difference between the states of the interacting particles. Similar to the preceding case, interactions can be dissipative or productive.
3. *Hiding — —chasing games* : One of the two attempts to increase the overall distance from the other, which attempts to reduce it. As a consequence, the distance can either increase or decrease.
4. *Games with Learning* : One of the two particles modifies, independently from the other, the microscopic state, in the sense that it learns by attempting to reduce the distance between them. In some cases, even the individual with higher level of knowledge can improve such a level by taking benefit of process of transfer of knowledge.
5. *Games with mutations* : Particle not only modifies their microscopic state, but also can move across functional subsystems.

Dynamics 1–4 are visualized in Fig. 3.2, while dynamics 5 is visualized later in Fig. 3.3.

Some general considerations are preliminary to the analytic treatment presented in the next paragraph.

- Particles, due to interactions, can modify both their microscopic states and move across functional subsystems. Namely, interactions may lead to mutations. In some cases, a new functional subsystem is generated.

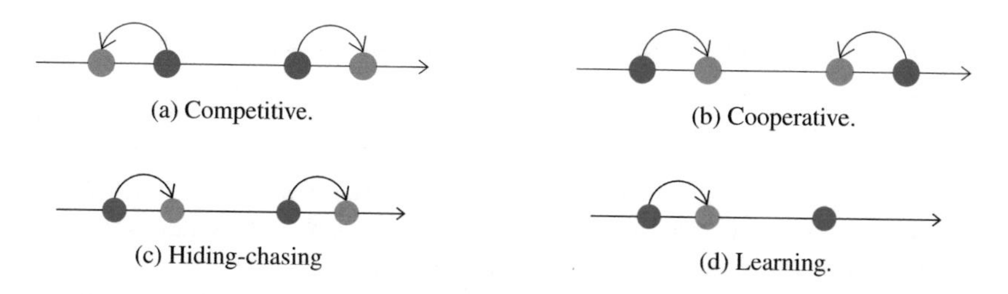

Fig. 3.2 Pictorial illustration of **a** competitive, **b** cooperative, **c** hiding-chasing, and **d** learning game dynamics between two active particles. *Blue* and *red circles* denote the pre- and post-interaction states of the particles, respectively

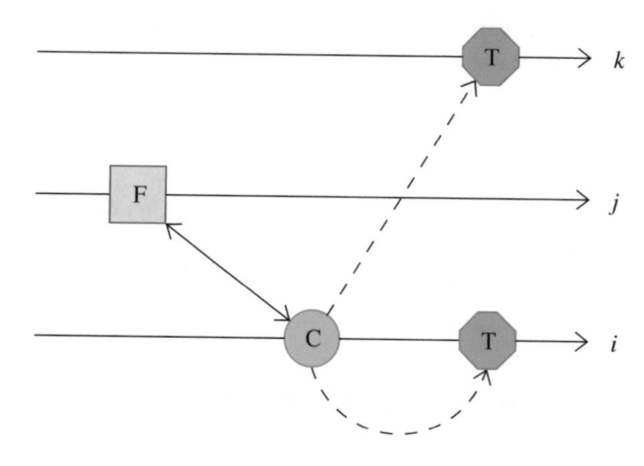

Fig. 3.3 Visualization of interactions. The candidate particle is denoted by the *light red circle*, the field particle by *light blue square*, and the test particle by *light green hexagon*. Notice that mutations can occur

- Interactions are not simply binary. For instance, an active particle can feel not only the action of the surrounding particles, but, in some cases, can feel also that of the whole system. As a possible example, a particle in a system with a space structure can be subject to an attraction toward the center of mass of the overall system.
- The term *linear interactions* is used when the output depends only on the microscopic state of the interacting particles, and the causality action is linearly additive. On the other hand, *nonlinear interactions* occur when the dynamics depends on the whole distribution. In both cases, interactions are not, generally, conservative. However, some quantities, such as the mean value or even higher-order moments, might be preserved in the output of the interaction.
- Competitive and cooperative interactions can be used to model some aspects of social dynamics, where the internal state is the wealth, while the cooperation is imposed by a taxation dynamics that impose to higher wealths to share part of their profits. Hiding–chasing interactions can be used in social systems such as the fight to reduce criminality, where detectives chase criminals approaching them, who try to escape. Learning dynamics and mutations are a feature of all types of interactions. Occasionally, the term *selfishness* is linked to competitive interactions, while *altruism* is linked to consensus interactions although, in some cases, this may correspond to an imitative trend that generates competition as it

occurs in panic conditions of crowds, when all pedestrians fight to follow the same path [2].

- Interactions present stochastic features as the output can be known only in probability; hence, the term *stochastic games* is used. All aforesaid types of games can occur simultaneously in a general context of heterogeneous particles. In some cases, see [46], the occurrence of one type with respect to the other is ruled by a *threshold* on the distance between the interacting active particles.
- The output of interactions can be determined by a collective strategy somehow different from that observed in the dynamics of a few entities. Namely, collective behaviors differ from that of a few entities, and such strategy can be modified by the environment. For instance, the presence of panic conditions induces different behaviors in various systems as observed in financial markets.
- In some cases, environmental conditions can modify both proliferative and destructive terms, for instance, higher proliferation for those individuals who are more fitted to the environment and higher death rate for those who are less fitted individuals.

The reader should be warned that the dynamics can even be more complex than that sketched above. The applications treated in the next chapters should clarify, at a practical level most of the issues that have been posed until now.

3.3.2 Modeling interactions by tools of game theory

This subsection aims at showing how the various interactions described at a qualitative level in the preceding subsection can be formally described by operators' modeling games. These can be used in the derivation of the aforementioned mathematical structure. As mentioned before, the case on networks will be studied in the next section, while games, as we shall see later, can modify the microscopic state of active particle also when they interact with agents which, conversely, do not modify their state.

Let us consider a large system constituted by m functional subsystems interacting in one node only. Each subsystem is described by the distribution function $f_i(t, u)$ with $i = 1, \dots, m$. The set of all $f_i = f_i(t, u)$ constitutes the dependent variables of the overall system whose evolution should be delivered by suitable models.

In principles, these stochastic variables need to be considered as distribution functions, rather than probability densities. In fact, the integral over the activity variable provides the size, namely the number of individuals, in each population of active particles. This number, due to different types of interactions and mortality, evolves in time.

The *dynamics of interactions*, which involve *candidate*, *field*, and *test* active particles, correspond to the various functional subsystems. In more detail

- **Test** particles of the ith functional subsystem with microscopic state, at time t, delivered by the activity variable u, whose distribution function is $f_i = f_i(t, u)$. The test particle is assumed to be representative of the whole system.
- **Field** particles of the kth functional subsystem with microscopic state, at time t, delivered by the activity variable u^*, whose distribution function is $f_k = f_k(t, u^*)$.
- **Candidate** particles, of the hth functional subsystem, with microscopic state, at time t, delivered by the activity variable u_*, whose distribution function is $f_h = f_h(t, u_*)$.

Figure 3.3 provides a pictorial representation of interactions, where particles can move across functional subsystems.

More in detail, we consider the following typology of interactions:

- *Number conservative interactions* : Candidate particles can acquire, in probability, the state of the test particles after interactions with field particles or with agents, while test particles lose their state after interaction with them;
- *Proliferative/destructive interactions* : Interactions induce a proliferative and/or destructive dynamics for active particles, which have a natural decay related to mortality;
- *Interactions with mutations* : Conservative interaction of active particles can induce a mutation of the particle into an other functional subsystem, or even a new one created by the mutating entity. Similarly, proliferative events of active particles can generate new particles in an other, or even a new, functional subsystem.

Let us now consider the various terms that can model dynamics of interactions at the microscopic scale, where particles belonging to the subsystem i are denoted, for short, by i-particles:

Interaction rate

- The interaction rates η_{hk} are positive definite quantities modeling the encounter rate of the interactions between h-particles and k-particles.

Number conservative interactions

- $\mathscr{A}^i_{hk}[\mathbf{f}](u_* \to u|u_*, u^*)$ models the probability density that a candidate h-particle, with state u_*, ends up into a i-particle with state u after the interaction with a field k-particle with state u^*. The terms $\mathscr{A}^i_{hk} \geq 0$ satisfy for all inputs the probability density property, i.e., normalization with respect to the output.

Proliferative/destructive interactions are modeled by the following probability distributions:

- $\mathscr{P}^{i+}_{hk}[\mathbf{f}](u_* \to u|u_*, u^*)$ models the birth process for a candidate h-particle, with state u_*, into the state u of the ith functional subsystem due the interaction with the field k-particles with state u^*.

• $\mathscr{P}^{i-}_{ik}[\mathbf{f}](u, u^*)$ models the destructive process for a test i-particles, with state u, into the same state and functional subsystem due the interaction with the field k-particles with state u^*.

Natural decay

• λ_i denote the coefficients of a natural trend toward equilibrium, or a death process of the i-particles due to their natural extinction, where death occurs only within the same functional subsystem and state of the candidate particles.

Remark 3.6 *All interaction terms can depend not only on the state of the interacting pairs, but also on their distribution functions. This feature introduces a nonlinearity of parameters in addition to the structural nonlinearity generated by the product of dependent variables. This delicate issue deserves a deep analysis related to modeling issues. Some hints are proposed in [36].*

Remark 3.7 *Interactions might be triggered by a critical value of a distance from interacting pairs, for instance separation between cooperation and competition. Recent papers [46, 107] suggest that the threshold is determined by the overall dynamics of the system and involves macroscopic scale quantities. In addition, it can be argued that in an heterogeneous system both dynamics coexist and that the critical value of the said distance models the prevalence of a dynamics over the others.*

3.3.3 Mathematical structures for closed systems

This subsection shows how a general framework suitable to provide the time–space evolution of the distribution functions can be derived based on the definition of the aforesaid interaction terms by a differential system for the dynamics of the probability distributions, which is obtained by a balance of particles within elementary volumes of the space of microscopic states, the inflow and outflow of particles being related to the aforementioned interactions.

More in detail, the balance of particles is as follows:

Variation rate of the number of active particles
= *Inlet flux rate caused by conservative interactions*
−*Outlet flux rate caused by conservative interactions*
+*Inlet flux rate caused by proliferative interactions*
−*Outlet flux rate caused by destructive interactions*
−*Natural trend toward an equilibrium distribution*,

where the inlet flux includes the dynamics of mutations.

Hence, technical calculations yield:

$$\partial_t f_i(t,u) = J_i[\mathbf{f}](t,u) = C_i[\mathbf{f}](t,u) + P_i[\mathbf{f}](t,u) - \lambda_i[f_i](f_i - f_{ie})(t,u)$$

$$= \sum_{h,k=1}^{m} \iint_{D_u \times D_u} \eta_{hk}[\mathbf{f}](u_*,u^*)\, \mathscr{A}_{hk}^{i}[\mathbf{f}](u_* \to u|u_*,u^*)\, f_h(t,u_*)\, f_k(t,u^*)\, du_*\, du^*$$

$$- f_i(t,u) \sum_{k=1}^{m} \int_{D_u} \eta_{ik}[\mathbf{f}](u,u^*)\, f_k(t,u^*)\, du^*$$

$$+ \sum_{h,k=1}^{m} \iint_{D_u \times D_u} \eta_{hk}[\mathbf{f}](u_*,u^*)\, \mathscr{P}_{hk}^{i+}[\mathbf{f}](u_* \to u|u_*,u^*)\, f_h(t,u_*)\, f_k(t,u^*)\, du_*\, du^*$$

$$- \sum_{k=1}^{m} \int_{D_u} \eta_{ik}[\mathbf{f}](u,u^*)\, \mathscr{P}_{ik}^{i-}[\mathbf{f}](u,u^*)\, f_i(t,u)\, f_k(t,u^*)\, du^*$$

$$-\lambda_i[f_i](f_i - f_{ie})(t,u), \tag{3.17}$$

where f_{ie} denotes the equilibrium distribution.

This structure simplifies when only conservative or proliferative dynamics occur.

Conservative interactions: Let us consider the dynamics of a system where only conservative interactions occur. Conservative terms can be viewed as the difference between a gain term of particles which acquire a new state in the ith functional subsystem due to interactions and a loss term of test particles that lose such state. Technical calculations related to the application of the balance of particles yield:

$$\partial_t f_i(t,u) = C_i[\mathbf{f}](t,u) - \lambda_i[f_i](f_i - f_{ie})(t,u)$$

$$= \sum_{h,k=1}^{m} \iint_{D_u \times D_u} \eta_{hk}[\mathbf{f}](u_*,u^*)\, \mathscr{A}_{hk}^{i}[\mathbf{f}](u_* \to u|u_*,u^*)\, f_h(t,u_*)\, f_k(t,u^*)\, du_*\, du^*$$

$$- f_i(t,u) \sum_{k=1}^{m} \int_{D_u} \eta_{ik}[\mathbf{f}](u,u^*)\, f_k(t,u^*)\, du^*$$

$$-\lambda_i[f_i](f_i - f_{ie})(t,u). \tag{3.18}$$

Proliferative/destructive interactions: Let us consider the dynamics of a system where only proliferative and destructive interactions occur. Also, in this case, the dynamics is represented by a difference between a "gain" and a "loss" terms, where the gain term is due to proliferation, while the loss term is due to destruction. Calculations analogous to those we have seen above yield:

$$
\begin{aligned}
\partial_t f_i(t,u) &= P_i^+[\mathbf{f}](t,u) - P_i^-[\mathbf{f}](t,u) - \lambda_i[f_i](f_i - f_{ie})(t,u)\\
&= \sum_{h,k=1}^{m} \iint_{D_u\times D_u} \eta_{hk}[\mathbf{f}](u_*,u^*)\, \mathscr{P}_{hk}^{i+}[\mathbf{f}](u_* \to u|u_*,u^*)\, f_h(t,u_*)\, f_k(t,u^*)\, du_*\, du^*\\
&\quad - \sum_{k=1}^{m} \int_{D_u} \eta_{ik}[\mathbf{f}](u,u^*)\, \mathscr{P}_{ik}^{i-}[\mathbf{f}](u,u^*)\, f_i(t,u)\, f_k(t,u^*)\, du^*\\
&\quad - \lambda_i[f_i](f_i - f_{ie})(t,u). \qquad (3.19)
\end{aligned}
$$

3.3.4 Mathematical structures for open systems

If the system is subject to external actions which are known functions of time and activity, the framework can be rapidly derived according to the same rationale. Let us first consider the case of a macroscopic external action $K_i(t,u)$ acting over each functional subsystem. The structure modifies as follows:

$$
\partial_t f_i(t,u) + \partial_u\big[K_i(t,u) f_i(t,u)\big] = J_i[\mathbf{f}], \qquad (3.20)
$$

while in a more general setting, the external action K_i can depend on f_i, namely $K_i[f_i](t,u)$.

External actions occur when particles of the inner system interact with agents of the outer system which is subdivided into the same subsystems. Therefore, external actions are, in this case, applied by agents with state $g_i(t,w)$ with $i = 1,\dots,m$, over the activity variable of each of them. Hence, the dependent variables of the overall system are the probability distributions f_i, as the g_i are known quantities.

The derivation of the mathematical structure needs introducing the following additional quantities:

- The interaction rates $\mu_{hk}[\mathbf{f},\mathbf{g}](u_*,w^*)$, which are positive defined quantities, modeling the encounter rates of the interactions between h-particles and k-agents.

- The conservative interaction terms $\mathscr{B}_{hk}^{i}[\mathbf{f},\mathbf{g}](u_* \to u|u_*,w^*)$, which models the probability density that a candidate h-particle, with state u_*, ends up into a i-particle with state u after the interaction with a field k-agent with state w^*. All terms $\mathscr{B}_{hk}^{i} \geq 0$ satisfy for all inputs the probability density property, i.e., normalization with respect to the output property.

- The proliferative terms $\mathscr{Q}_{hk}^{i+}[\mathbf{f},\mathbf{g}](u_* \to u|u_*,w^*)$ which model the birth process for a candidate h-particle, with state u_*, into the state u of the ith functional subsystem due to the interaction with the field k-agents with state w^*.

• The destructive terms $\mathscr{Q}_{ik}^{i-}[f_i, \mathbf{g}](u, w^*)$ which model the destructive process for a test i-particle, with state u, into the same state and functional subsystem due the interaction with the field k-agent with state w^*.

Implementing the dynamics modeled by these terms into the structure derived in Subsection 3.3.3 by the same balance of particles in the elementary volume of the microscopic states provides a general mathematical structure which include the action, supposed known, of the external agents, as follows:

$$\begin{aligned}
\partial_t f_i(t,u) &= C_i[\mathbf{f}](t,u) + C_i^e[\mathbf{f},\mathbf{g}](t,u) + P_i[\mathbf{f}](t,u) + P_i^e[\mathbf{f},\mathbf{g}](t,u)\\
&\quad - D_i[\mathbf{f}](t,u) - D_i^e[\mathbf{f},\mathbf{g}](t,u) - \lambda_i[f_i](f_i - f_{ie})(t,u)\\
&= \sum_{h,k=1}^{m} \iint_{D_u\times D_u} \eta_{hk}[\mathbf{f}](u_*,u^*)\, \mathscr{A}_{hk}^{i}[\mathbf{f}](u_* \to u|u_*,u^*)\, f_h(t,u_*)\, f_k(t,u^*)\, du_*\, du^*\\
&\quad - f_i(t,u) \sum_{k=1}^{m} \int_{D_u} \eta_{ik}[\mathbf{f}](u,u^*)\, f_k(t,u^*)\, du^*\\
&\quad + \sum_{h,k=1}^{m} \iint_{D_u\times D_u} \eta_{hk}[\mathbf{f}](u_*,u^*)\, \mathscr{P}_{hk}^{i+}[\mathbf{f}](u_* \to u|u_*,u^*)\, f_h(t,u_*)\, f_k(t,u^*)\, du_*\, du^*\\
&\quad - \sum_{k=1}^{m} \int_{D_u} \eta_{ik}[\mathbf{f}](u,u^*)\, \mathscr{P}_{ik}^{i-}[\mathbf{f}](u,u^*)\, f_i(t,u)\, f_k(t,u^*)\, du^*\\
&\quad + \sum_{h,k=1}^{m} \iint_{D_u\times D_w} \mu_{hk}[\mathbf{f},\mathbf{g}](u_*,w^*)\, \mathscr{B}_{hk}^{i}[\mathbf{f},\mathbf{g}](u_* \to u|u_*,w^*)\\
&\qquad \times f_h(t,u_*)\, g_k(t,w^*)\, du_*\, dw^*\\
&\quad - f_i(t,u) \sum_{k=1}^{m} \int_{D_w} \mu_{ik}[\mathbf{f},\mathbf{g}](u,w^*)\, g_k(t,w^*)\, dw^*\\
&\quad + \sum_{h,k=1}^{m} \iint_{D_u\times D_w} \mu_{hk}[\mathbf{f},\mathbf{g}](u_*,w^*)\, \mathscr{Q}_{hk}^{i+}[\mathbf{f},\mathbf{g}](u_* \to u|u_*,w^*)\\
&\qquad \times f_h(t,u_*)\, g_k(t,w^*)\, du_*\, dw^*\\
&\quad - \sum_{k=1}^{m} \int_{D_w} \mu_{ik}[\mathbf{f},\mathbf{g}](u,w^*)\, \mathscr{Q}_{ik}^{i-}[f_i,\mathbf{g}](u,w^*) f_i(t,u)\, g_k(t,w^*)\, dw^*\\
&\quad - \lambda_i[f_i](f_i - f_{ie})(t,u). \qquad (3.21)
\end{aligned}$$

Remark 3.8 *These structures put in evidence two types of nonlinearities: the quadratic one given by the product of dependent variables* $\mathbf{f}$ *and the nonlinearity of the* *parameters modeling interactions, namely η, μ, $\mathscr{A}$, $\mathscr{B}$, $\mathscr{P}$, and $\mathscr{Q}$, which formally involve the dependent variable* $\mathbf{f}$.

3.3.5 Sources of nonlinearity

The various terms modeling interactions are allowed, as mentioned before, to depend on the distribution functions of the active particles. In fact, both the interaction rates and the output of the game can depend not only on the microscopic state of the interacting particles, but also on the set of distribution functions, for instance on moments computed for the particles within the interaction domain, say mean value or low-order moments. Analogous remark can be addressed to proliferative and destructive interactions.

A detailed overview of these nonlinearities can be useful toward a deep understanding of the dynamics of interactions. The analysis, presented in the following, refers to interactions between h-candidate and k-field particles, namely belonging to the h and k functional subsystems, while their states are, respectively, u_* and u^*. The specific quantities that depend on the probability distributions and that can have an influence on the modeling of the said interaction terms are examined in the following, while the next chapters will put in evidence their role at a practical level.

Sensitivity domain: Interactions depend also on the domain of the activity variable, within which each particle has the ability to perceive a sufficient amount of signals and develop consequently a strategy. Consider then an active particle interacting with particles in a subdomain Ω of D_u. A sufficient information is achieved if interaction involves a number n_c of field particles. According to [52], one has to distinguish a *sensitivity domain* Ω_S where a particle *can* feel other particles and the influence domain Ω_I needed for a sufficient information and used to organize the dynamics.

If $\Omega_I \subseteq \Omega_S$, the particle can elaborate its strategy, while if $\Omega_S \subseteq \Omega_I$, the particle does not receive the said information and may produce an irrational strategy. Namely, a candidate (or test) particle interacts with a number of field particles by means of a communication ability that is effective only within a certain *domain of influence* of $\Omega[\mathbf{f}]$, which depends on the maximal density of active particles which can be captured in the communication. This domain is effective only if it is included in the *sensitivity domain* Ω_S, within which active particles have the potential ability to feel the presence of another particles. The relation between n_c and Ω_I is, for a generic f, as follows:

$$n_c = \int_{\Omega_I[f;n_c]} f(t,u)\,du, \tag{3.22}$$

which has a unique solution only in some special cases. For instance, if u is a scalar defined over the whole real axis, and the sensitivity is symmetric with respect to u, one has:

$$n_c = \int_{u-s[f;n_c]}^{u+s[f;n_c]} f(t,u)\,du, \tag{3.23}$$

which allows to compute $s[f; n_c]$, so that

$$\Omega_I = [u - s[f; n_c], u + s[f; n_c]]$$

is defined. On the other hand, if u is defined in a bounded domain or the sensitivity is not symmetrical, additional assumptions are needed to compute Ω_I.

Distance between interacting particles: The following different concepts of distance can play an important role in the interaction dynamics. Notations are referred to the active particles, namely to the inner system; however, the generalization to agents, namely to the external system, is immediate.

(i) *State distance*, which simply depends on the distance $|u_* - u^*|$ between the microscopic states of the interacting entities.
(ii) *Individual-mean state distance*, which provides a different interpretation of the previous one and refers the state u_* to the mean value of u^* in the domain of interactions of the field particles according to a suitable metric which can be formally denoted as follows: $|u_* - \mathbb{E}_\Omega(u^*)|$, where $\mathbb{E}_\Omega(u^*)$ denotes the mean value of u^* in the domain Ω.
(iii) *Hierarchic distance*, which occurs when two active particles belong to different functional subsystems. Then, the distance $|k - h|$ can be defined if a conceivable numbering criterion is applied in selecting the first subsystem by a certain selection rule (for instance, in the animal world, the "dominant") and in numbering the others by increasing numbers depending on the decreasing rate. Namely, $|k - h| \uparrow \Rightarrow \eta_{hk} \downarrow$.
(iv) *Affinity distance*, when the distance between the distribution functions of interacting active particles can be defined according to the general idea that two systems with close distributions are *affine*. In this case, the distance is $||f_h - f_k||_{L^p(\Omega[\mathbf{f}])}$, where a uniform, e.g., $L_\infty(\Omega[\mathbf{f}])$, or mean, e.g., $L_1(\Omega[\mathbf{f}])$, approximation can be adopted as the most appropriate according to the physics of the system under consideration. As above, the rate decays with increasing distance.

Remark 3.9 *A very simple metrics has been used. On the other hand, considering that the distance involves probability distributions, the use of more sophisticated concepts of distances should be studied. For instance, the use of Wasserstein [239] type metrics should be properly investigated.*

Of course, only the distances defined in (ii) and (iv) induce nonlinearities, while the state distance (i) refers to the microscopic state and hence does not generate a nonlinear term.

The hierarchic distance appears when individuals of a population can determine the dynamics of other individual. It is a delicate matter, which requires a detailed analysis of the system which is being modeled. In some cases, a hierarchy is an inner feature of the system, while in other cases is a consequence of the dynamics.

These concepts can be used toward the modeling of the interaction rates, for instance based on the idea that the interaction rate decays with the distance. As an example, the interaction rate between active candidate and field particles can be modeled by using the following general structure:

$$\eta_{hk}[\mathbf{f}] = \eta^0 \exp\big\{ - d_{hk}[\mathbf{f}](u_*, u^*)\big\}, \tag{3.24}$$

where

$$d_{hk}[\mathbf{f}] = c_1|u_* - u^*| + c_2|h - k| + c_3|u_* - \mathbb{E}_\Omega(u^*)| + c_4||f_h - f_k||_{L^p(\Omega_l)}. \tag{3.25}$$

Critical size toward survival of functional subsystems: The survival of functional subsystems might, in some cases, depend on their size. If the size $n_i[f_i]$ of a functional subsystem falls below a critical value N_m, then interactions reduce to transfer individuals from the original one toward an aggregation to another one. In principles, a large critical size N_m might exist such that if the said threshold is overcome, particles are induced to move into another functional subsystem. This type of dynamics occurs in various social systems, where functional subsystems are identified by groups of interest [4]; for instance, in the case of voting dynamics, the critical size is a minimal number of entities that assure the survival of the party; when this size is not reached, supporters move from that party to an other.

3.4 Mathematical Structures for the Dynamics in a Network

This subsection shows how the derivation of the said mathematical structure can be obtained also in the case of a network. The theoretical tools of games theory are used also in this case, where the only difference is that interactions occur also between entities of different nodes. Similar rules are followed in the case of interactions with external agents, see [163], when the *inner system* interacts with the *outer system* made of agents, whose action is known in time.

The derivation is not developed in the most general case, as one should also consider the case of presence in the same node of more than one functional subsystem, proliferative/destructive interactions, as well the case of dynamics of the external agents. In fact, the mathematical structure is derived only in the case of number conservative interactions. On the other hand, further generalizations, although technical, would require heavy notations. Thus, it appears more appropriate dealing with them referring to specific applications. In fact, sufficient information are, however, given here.

The main features of the interactions that a candidate (or test) particle can undergo are sketched, still according to [163], in the following chart:

•Within the same node	(i) Binary interactions with field particles within the same functional subsystem (ii) Interactions with the mean activity within each functional subsystem
•Within the network	(iii) Interactions with the mean activity of the other functional subsystems (iv) Interactions with the mean activity of the whole network
•With the environment	(v) Binary interactions with agents of the outer environment (vi) Interactions with the mean activity of the external agents

Let us now consider, for each type of interactions, the formal structure of the terms which model them. We consider, in the following, a scalar activity variable and one only functional subsystem in each node. Therefore, the node coincides with the functional subsystem. Moreover, as already mentioned, the derivation is limited to the dynamics of the inner system only, while the external actions are supposed to be given. The terms modeling interactions, where FS abbreviates "functional subsystem," are the following:

(i) Binary interactions within the same node (functional subsystem):

- $\eta_i[f_i](u_*, u^*)$ denotes the *encounter rate* between a candidate particle with activity u_* and a field particle with activity u^*, both within the i-FS. This rate can nonlinearly depend on f_i.
- $\mathscr{B}_i[f_i](u_* \to u|u_*, u^*)$ models the *probability density* that a candidate particle with state u_* falls into the state u after the interaction with a field particle with state u^*. Interactions can be nonlinearly additive, namely are influenced by the dependent variable f_i.

(ii) Interactions with the mean activity $\mathbb{E}_i$ within the same FS:

- $\mu_i[f_i](u_*, \mathbb{E}_i)$ denotes the *interaction rate* between a candidate particle with state u_* and the mean value $\mathbb{E}_i$. This term nonlinearly depend on f_i.
- $\mathscr{M}_i[f_i](u_* \to u|u_*, \mathbb{E}_i)$ denotes the *probability density* that a candidate particle with state u_* ends up into the state u, within the same FS, after interaction with the mean activity value $\mathbb{E}_i$.

(iii) Interactions of particles of the h-FS with the mean activity of other k-FSs:

- $\mu_{hk}[f_h, f_k](u_*, \mathbb{E}_k)$ is the *interaction rate*, which can nonlinearly depend on the distribution functions of the interacting functional subsystems corresponding to the interacting nodes h and k.
- $\mathscr{M}^i_{hk}[f_h, f_k](u_* \to u, h \to i|u_*, \mathbb{E}_k)$ denotes the *probability density* that a candidate particle belonging to h-FS with state u_* ends up in the state u of the i-FS after interaction with the mean activity value of the k-FS.

(iv) Interactions of particles of the i-FS with the mean activity $\mathbb{E}_{net}$ of the whole network:

- $\mu_{iN}[\mathbf{f}](u_*, \mathbb{E}_{net})$ is *interaction rate*, where square brackets over $\mathbf{f}$ denote the dependence on the whole set of distribution functions.
- $\mathcal{M}_{iN}[\mathbf{f}](u_* \to u|u_*, \mathbb{E}_{net})$ models change of microscopic scale state within the same FS.

Let us now consider the dynamics of the inner system as subject to interactions with the outer system. The following terms can model interactions:
(v) Binary interactions with external agents: Interactions with the outer environment are represented by binary interactions between (candidate) active particles and (field) external agents.

- $\eta_i^e[f_i, g_i](u_*, w^*)$ models the *interaction rate* of binary interactions between a candidate particle belonging to i-FS with activity u_* and the ith external agent with activity u^*.
- $\mathcal{B}_i^e[f_i, g_i](u_* \to u|u_*, w^*)$ models the *probability density* that a candidate particle belonging to i-FS with state u_* ends up into the state u after an interaction with an external agent with state w^*.

(vi) Interactions with the mean activity of the external agent: This kind of interactions takes place when the candidate particle of the ith FS takes into account the mean activity value of the external agents g_i, denoted by $\mathbb{E}_i^e$.

- $\mu_i^e[f_i, g_i](u_*, \mathbb{E}_i^e)$ defines the *interaction rate* between a candidate particle with state u_* and the mean external value given by $\mathbb{E}_i^e$.
- $\mathcal{M}_i^e[f_i, g_i](u_* \to u|u_*, \mathbb{E}_i^e)$ models the *probability density* that a candidate particle with state u_* ends up into the state u after interacting with the mean external value $\mathbb{E}_i^e$.

Similar to the approach related to one node only, the terms $\mathcal{B}_i, \mathcal{M}_i, \mathcal{M}_{hk}^i, \mathcal{M}_{iN}, \mathcal{B}_i^e$, $\mathcal{M}_i^e$, are probability densities; hence, their integral over the output of the interactions is equal to one for all inputs, namely their arguments. In addition, considering that we have not included proliferative and destructive interactions, this specific structure is conservative in the whole network, namely:

$$\sum_{i=1}^{n} n_i[f_i](t) = N_0. \tag{3.26}$$

Technical calculations analogous to those of the preceding section yield:

$$\partial_t f_i(t,u) = J_i[f_i](t,u) + M_i[\mathbf{f}](t,u) + N_i[\mathbf{f}](t,u) + N_i^N[\mathbf{f}](t,u)$$
$$+ Q_i[\mathbf{f},\mathbf{g}](t,u) + R_i[\mathbf{f},\mathbf{g}](t,u) - \lambda_i[f_i](f_i - f_{ie})(t,u), \tag{3.27}$$

where:

$$J_i = \int_{D_u \times D_u} \eta_i[f_i](u_*, u^*)\mathscr{B}_i(u_* \to u|u_*, u^*) f_i(t, u_*) f_i(t, u^*) du_* du^*$$
$$- f_i(t, u) \int_{D_u} \eta_i[f_i](u, u^*) f_i(t, u^*) du^*, \tag{3.28}$$

$$M_i = \int_{D_u} \mu_i[f_i](u_*, \mathbb{E}_i)\mathscr{M}_i[f_i](u_* \to u|u_*, \mathbb{E}_i) f_i(t, u_*) du_*$$
$$- \mu_i[f_i](u, \mathbb{E}_i) f_i(t, u), \tag{3.29}$$

$$N_i = \sum_{\substack{h,k=1 \\ h \neq k}}^{n} \int_{D_u} \mu_{hk}[f_h, f_k](u_*, \mathbb{E}_k)\mathscr{M}_{hk}^i[f_h, f_k](u_* \to u, h \to i|u_*, \mathbb{E}_k) f_h(t, u_*) du_*$$
$$- f_i(t, u) \sum_{\substack{k=1 \\ k \neq i}}^{n} \mu_{ik}[f_i, f_k](u, \mathbb{E}_k), \tag{3.30}$$

$$N_i^N = \int_{D_u} \mu_{iN}[\mathbf{f}](u_*, \mathbb{E}_{net})\mathscr{M}_{iN}[\mathbf{f}](u_* \to u|u_*, \mathbb{E}_{net}) f_i(t, u_*) du_*$$
$$- \mu_{iN}[\mathbf{f}](u, \mathbb{E}_{net}) f_i(t, u), \tag{3.31}$$

$$Q_i = \int_{D_u \times D_w} \eta_i^e[f_i, g_i](u_*, w^*)\mathscr{B}_i^e[f_i, g_i](u_* \to u|u_*, w^*) f_i(t, u_*) g_i(t, w^*) du_* dw^*$$
$$- f_i(t, u) \int_{D_w} \eta_i^e[f_i, g_i](u, w^*) g_i(t, w^*) dw^*, \tag{3.32}$$

and

$$R_i = \int_{D_u} \mu_i^e[f_i, g_i](u_*, \mathbb{E}_i^e)\mathscr{M}_i^e[f_i, g_i](u_* \to u|u_*, \mathbb{E}_i^e) f_i(t, u_*) du_*$$
$$- \mu_i^e[f_i, g_i](u, \mathbb{E}_i^e) f_i(t, u). \tag{3.33}$$

Remark 3.10 *Although the structure presented in this subsection can include some specific features of networks, it is not naively claimed that the topic is exhaustively treated. The literature in the field is rapidly growing with a blow up of interesting contributions. Some titles, among several ones, are brought to the attention of the interested reader toward a deeper understanding of this challenging topic [8, 21, 22, 24, 93, 144, 228, 236]. Additional bibliography will be given referring to specific applications.*

3.4.1 Structures for discrete activity variables

Various applications in different fields, such as the modeling of social systems [46] or of the immune competition [57], suggest the use of discrete activity variables. In fact, it is difficult, in several cases obtaining by empirical data a precise assessment of the probability distribution over the activity, while it is practical counting the active particles within a certain subset of the domain of such a variable.

Let us now rapidly show how the derivation of the mathematical structure can be obtained in the very simple case of a system in a single node, undergoing simply conservative interactions, in absence of external actions.

Therefore, let us consider the system of Subsection 3.3.2, where in addition to the subdivision into functional subsystems, labeled by the subscript $i = 1, \dots, n$, a discrete activity variable, labeled by the subscript $j = 1, \dots, m$ with $u_1 = 0$ and $u_m = 1$, is introduced for each of them as by Remark 3.3. Hence, the domain of the activity variable has been normalized in the interval [0, 1]. The abbreviation ij-particle is used for a particle belonging to the ith functional subsystem with activity u_j. The representation of the overall system is delivered by the discrete probability

$$f_{ij} = f_{ij}(t) \, : \, [0, T] \to \mathbb{R}_+, \tag{3.34}$$

while weighted sums replace the integrals to obtain variables at the macroscopic scale.

The actual derivation of the new mathematical structure needs defining the interaction terms in the new context. In detail:

- η_{hk}^{pq} is the encounter rate between a candidate hp-particle and a field kq-particle.
- μ_{hp} is the encounter rate between a candidate hp-particle and the mean activity within its functional subsystem.
- $\mathscr{A}\mathscr{D}_{hk}^{pq}(ij)$ is the discrete transition probability that a candidate hp-particle ends up into the state of the test ij-particle due the interaction with a field kq-particle.
- $\mathscr{M}\mathscr{D}_h^p(u_p, \mathbb{E}_h)(ij)$ is the transition probability that a candidate hp-particle ends up into the state ij-particle due the interaction with the mean activity value $\mathbb{E}_h$.

The transition probabilities satisfy, for all type of inputs, the following normalization

$$\sum_{i=1}^{n} \sum_{j=1}^{m} \mathscr{A}\mathscr{D}_{hk}^{pq}(ij) = \sum_{i=1}^{n} \sum_{j=1}^{m} \mathscr{M}\mathscr{D}_h^p(u_p, \mathbb{E}_h)(ij) = 1. \tag{3.35}$$

The balance of particles in the elementary volume of the space of microscopic states leads to the following structure:

$$\partial_t f_{ij}(t) = \sum_{h,k=1}^{n} \sum_{p,q=1}^{m} \eta_{hk}^{pq} \mathscr{A}\mathscr{D}_{hk}^{pq}(ij) f_{hp}(t) f_{kq}(t)$$

$$- f_{ij}(t) \sum_{k=1}^{n} \sum_{q=1}^{m} \eta_{ik}^{jq} f_{kq}(t)$$

$$+ \sum_{h=1}^{n} \sum_{p=1}^{m} \mu_{hp} \mathscr{M}\mathscr{D}_{h}^{p}(ij) f_{hp}(t)$$

$$- \mu_{ij}(u_j, \mathbb{E}_i) f_{ij}(t). \tag{3.36}$$

Equation (3.36) provides a simple example of a possible discrete mathematical structures derived looking at specific applications, where only conservative interactions have been taken into account. Generalizations to proliferative and destructive dynamics are simply a matter of technical calculations.

3.5 Structures when Space is a Continuous Variable

This section is devoted to the derivation of a general framework for systems with a space structure. Then, a large system of interacting functional subsystems in a domain $\Sigma \subseteq \mathbb{R}^3$ is considered. The derivation of such a structure follows the same rationale presented in preceding section. However, some important modifications have to be considered to model the role of space both on the transport term on the left-hand side of the equality sign and on the interaction operator on the right-hand side.

The presentation is proposed along the following steps treated in the next subsections:

1. Representation;
2. Modeling interactions;
3. Derivation of a mathematical structure;
4. Mean field models; and
5. Perturbation of space homogeneity.

The presentation will account for all technical modifications needed to generalize the calculations proposed in Section 3.3. For sake of completeness, the calculations are entirely repeated and include the aforementioned modifications.

3.5.1 Representation

Let us consider a system of active particles, whose microscopic state includes geometrical and mechanical variables, for instance position and velocity, and a scalar activity variable. These particles interact within a domain $\Sigma \subseteq \mathbb{R}^3$ of the space variable. If the domain is bounded, the largest dimension is denoted by ℓ. In some cases, the dynamics occurs in unbounded domains, and then, a reference length ℓ is still needed by reasoning on the physics and geometry of the system under consideration.

If particles are modeled as points, their microscopic state is identified by the following dimensionless variables:

- $\mathbf{x} = (x, y, z)$ is the position, referred to ℓ, of each particle;
- $\mathbf{v} \in D_\mathbf{v}$ is the velocity, whose modulus is referred to the maximal velocity v_M, which particles can attain for their own physical limit;
- $u \in D_u$ is the activity variable, which models the strategy developed by particles; and
- The microscopic state is the set of all the variables $\mathbf{x}$, $\mathbf{v}$, u, and it defines the state of each individual.

Following the same hallmarks of the preceding section, particles are subdivided into a number "m" of *functional subsystems* corresponding to groups of particles that express the same activity. In some cases, although the activity is the same, a different way to express it can characterize different subsystems.

The overall state is described by the *distribution functions* over the microscopic state of the interacting particles

$$f_i = f_i(t, \mathbf{x}, \mathbf{v}, u) : \quad [0, T] \times \Sigma \times D_\mathbf{v} \times D_u \to \mathbb{R}_+ , \quad i = 1, \ldots, m, \tag{3.37}$$

where i denotes the functional subsystem.

The distribution functions are positive defined and are referred to the total number of particles N_0 at $t = 0$. Under suitable local integrability assumptions, it provides via $f_i(t, \mathbf{x}, \mathbf{v}, u)\, d\mathbf{x}\, d\mathbf{v}\, du$ the number of active particles that, for each functional subsystem at time t, are in the elementary volume of the space of the microscopic states

$$[\mathbf{x}, \mathbf{x} + d\mathbf{x}] \times [\mathbf{v}, \mathbf{v} + d\mathbf{v}] \times [u, u + du].$$

If the distribution functions f_i are known, macroscopic quantities can be computed as weighted moments by standard calculations analogous to those of Section 3.2. For instance, zeroth-order moments correspond to number density, while first- and second-order moments to quantities that can be interpreted as somehow equivalent to linear momentum and energy, respectively. Let us first show how mechanical quantities can be computed by moments weighted by the velocity variable and averaged over velocity and activity variables.

In more detail, the *local densities* for each functional subsystem are computed as follows:

$$\rho_i[f_i](t, \mathbf{x}) = \int_{D_\mathbf{v} \times D_u} f_i(t, \mathbf{x}, \mathbf{v}, u)\, d\mathbf{v}\, du\,. \tag{3.38}$$

while the *local mean velocity* is given by

$$\boldsymbol{\xi}_i[f_i](t, \mathbf{x}) = \frac{1}{\rho_i[f_i](t, \mathbf{x})} \int_{D_\mathbf{v} \times D_u} \mathbf{v}\, f_i(t, \mathbf{x}, \mathbf{v}, u)\, d\mathbf{v}\, du, \tag{3.39}$$

that defines the flow

$$\mathbf{q}_i[f_i](t, \mathbf{x}) = \int_{D_\mathbf{v} \times D_u} \mathbf{v}\, f_i(t, \mathbf{x}, \mathbf{v}, u)\, d\mathbf{v}\, du. \tag{3.40}$$

Second-order moments give the *local kinetic energy*

$$\mathscr{E}_i[f_i](t, \mathbf{x}) = \int_{D_\mathbf{v} \times D_u} \frac{1}{2} m v^2\, f_i(t, \mathbf{x}, \mathbf{v}, u)\, d\mathbf{v}\, du, \tag{3.41}$$

where m is the mass of the particles.

Social quantities are computed by moments weighted by the activity variable. In particular, the *local activation* is computed as follows:

$$a_i[f_i](t, \mathbf{x}) = \int_{D_\mathbf{v} \times D_u} u f_i(t, \mathbf{x}, \mathbf{v}, u)\, d\mathbf{v}\, du, \tag{3.42}$$

while the *local activation density* is given by

$$\alpha_i[f_i](t, \mathbf{x}) = \frac{a_i[f_i](t, \mathbf{x})}{\rho_i[f_i](t, \mathbf{x})} = \frac{1}{\rho_i[f_i](t, \mathbf{x})} \int_{D_\mathbf{v} \times D_u} u f_i(t, \mathbf{x}, \mathbf{v}, u)\, d\mathbf{v}\, du. \tag{3.43}$$

In addition, it might be useful looking at marginal densities, which correspond either to the activity variable, after integration over the mechanical variables, or to mechanical quantities after averaging over the activity variable. In detail:

$$f_i(t, \mathbf{x}, u) = \int_{D_\mathbf{v}} f_i(t, \mathbf{x}, \mathbf{v}, u)\, d\mathbf{v}, \tag{3.44}$$

$$f_i(t, \mathbf{v}, u) = \int_{\Sigma} f_i(t, \mathbf{x}, \mathbf{v}, u)\, d\mathbf{x}, \tag{3.45}$$

and

$$f_i(t, \mathbf{x}, \mathbf{v}) = \int_{D_u} f_i(t, \mathbf{x}, \mathbf{v}, u)\, du. \tag{3.46}$$

The distribution over the activity variable only is obtained by integration over space and velocity

$$f_i(t, u) = \int_{\Sigma \times D_{\mathbf{v}}} f_i(t, \mathbf{x}, \mathbf{v}, u)\, d\mathbf{x}\, d\mathbf{v}, \tag{3.47}$$

while global quantities are obtained by integration over the activity domain.

If particles have a shape to be considered in the modeling of the dynamics, then additional variables related to position and velocity are needed to take into account this feature, specifically angular variables and rotations. However, the representation does not include the treatment of entities with a variable shape due to an inner dynamics. This important issue needs additional analysis which is not treated in this book.

3.5.2 Modeling interactions

Modeling interactions follow the same rationale of the spatially homogeneous case treated in Section 3.3. However, some important modifications are required to account for the role of the space and velocity variables.

Active particles *play a game* at each interaction with an output that depends on their strategy which are often related to surviving and adaptation abilities, namely to an individual or collective search for fitness [52]. The output is not deterministic even when a causality principle is identified. This dynamics is also related to the fact that agents receive a feedback from their environment, which modifies the strategy they express adapting it to the mutated environmental conditions [173, 174], and hence, the dynamics of interactions evolves in time.

The following active particles are involved, for each functional subsystem, in the interactions:

- **Test** particles of the ith functional subsystem with microscopic state, at time t, delivered by the variable $(\mathbf{x}, \mathbf{v}, u)$, whose distribution function is $f_i = f_i(t, \mathbf{x}, \mathbf{v}, u)$. The test particle is assumed to be representative of the whole system.

- **Field** particles of the kth functional subsystem with microscopic state, at time t, defined by the variable $(\mathbf{x}^*, \mathbf{v}^*, u^*)$, whose distribution function is $f_k = f_k(t, \mathbf{x}^*, \mathbf{v}^*, u^*)$.

- **Candidate** particles, of the hth functional subsystem, with microscopic state, at time t, defined by the variable $(\mathbf{x}_*, \mathbf{v}_*, u_*)$, whose distribution function is $f_h = f_h(t, \mathbf{x}_*, \mathbf{v}_*, u_*)$.

Let us now consider short-range interactions, when particles interact within an interaction domain $\Omega \subset \Sigma$, generally small with respect to Σ. Bearing in mind that a precise definition and computing of Ω still needs to be given, the modeling of interactions at the microscopic scale can be described by the following quantities, where the term i-particle is used to denote a particle in the ith functional subsystem.

Similar to the preceding section, interactions are modeled by the following terms:

- *Interaction rate*, denoted by $\eta_{hk}[\mathbf{f}](\mathbf{x}, \mathbf{v}_*, u_*, \mathbf{x}^*, \mathbf{v}^*, u^*)$, which models the frequency of the interactions between a candidate h-particle with state $\mathbf{x}, \mathbf{v}_*, u_*$ and a field k-particle with state $\mathbf{x}^*, \mathbf{v}^*, u^*$. Analogous expression is used for interactions between test and field particles.
- *Transition probability density* $\mathscr{A}^i_{hk}[\mathbf{f}](\mathbf{x}, \mathbf{v}_*, u_* \to \mathbf{x}, \mathbf{v}, u|\mathbf{v}_*, u_*, \mathbf{v}^*, u^*)$, which denotes the probability density that a candidate h-particle, with state $\mathbf{x}, \mathbf{v}_*, u_*$, ends up into the state of the test particle of the ith functional subsystem after an interaction with a field k-particle.
- *Proliferative term* $\mathscr{P}^i_{hk}[\mathbf{f}](\mathbf{x}, \mathbf{v}_*, u_* \to \mathbf{x}, \mathbf{v}, u|\mathbf{v}_*, u_*, \mathbf{v}^*, u^*)$, which models the proliferative events for a candidate h-particle, with state $\mathbf{x}, \mathbf{v}_*, u_*$, into the ith functional subsystem after interaction with a field k-particle with state $\mathbf{x}^*, \mathbf{v}^*, u^*$.
- *Destructive term* $\mathscr{D}_{ik}[\mathbf{f}](\mathbf{x}, \mathbf{v}, u, \mathbf{v}^*, u^*)$, which models the rate of destruction for a test i-particle in its own functional subsystem after an interaction with a field k-particle with state $\mathbf{x}, \mathbf{v}^*, u^*$.

Remark 3.11 *A commonly applied assumption is that the terms* $\mathscr{P}^i_{hk}$ *and* $\mathscr{D}_{ik}$ *depend, in addition to* $\mathbf{f}$*, only on the activity variables, namely* $\mathscr{P}^i_{hk}[\mathbf{f}](u_* \to u|u_*, u^*)$ *and* $\mathscr{D}_{ik}[\mathbf{f}](u, u^*)$*.*

These quantities can be viewed in terms of rates by multiplying their interaction rate with the terms modeling transition, proliferative, and destructive events. Therefore, one has

- *Transition rate* : $\eta_{hk}[\mathbf{f}]\, \mathscr{A}^i_{hk}[\mathbf{f}]$;
- *Proliferation rate* : $\eta_{hk}[\mathbf{f}]\, \mathscr{P}^i_{hk}[\mathbf{f}]$;
- *Destruction rate* : $\eta_{ik}[\mathbf{f}]\, \mathscr{D}_{ik}[\mathbf{f}]$.

Remark 3.12 *Interactions are, as in the spatially homogeneous case, nonlinearly additive and non-local. The output modifies both velocity and activity locally for candidate and test particles.*

3.5.3 Preliminary rationale toward modeling interactions

Modeling interactions follow the same rationale of the spatially homogeneous case. However, some important modifications are required to account for the role of space and velocity variable. In practice the modeling can take advantage of suitable elaboration of the concepts of interaction domains and of distance between particles for the encounter rate, and of game theory for the transition probability density, while proliferative and destructive terms occur with the said encounter rate with an intensity

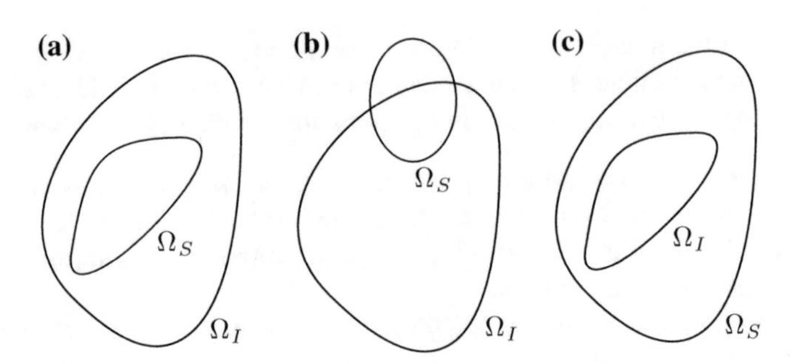

Fig. 3.4 Interplay between visibility and sensitivity zones

depending on the properties, namely state and functional subsystem, of the interacting active particles. The elaboration, with respect to the contents of Section 3.3, mainly refers to the additional technical problems induced by the space variable.

Sensitivity and space interaction domains: A candidate (or test) particle interacts with a number of field particles by means of a communication ability that is effective only within a certain *domain of influence* $\Omega_I[\mathbf{f}]$ of the space variable which depends on the maximal density of active particles which can be captured in the communication. This domain is effective only if it is included in the *sensitivity domain* $\Omega_S(\mathbf{x})$, within which active particles have the potential ability to feel the presence of another particles. In many cases, the sensitivity domain corresponds to a visibility zone that might be reduced by obstacles (geometry) or environmental (lack of visibility) reasons, but the sensitivity domain could also depend on many different other signals or waves such us sounds or other type of perceptions. Therefore, Ω_S depends on the localization of the particle, see Fig. 3.4.

Following [52], detailed calculations can be developed when the geometry of the sensitivity shape is known. For instance, it can be a circle or an arc of circle centered in the candidate (or test) particle as shown in Fig. 3.5. In this case, the domain of influence is simply identified by the radius of the geometry and by the visibility angle. Then, the conjecture on the critical density [19] can be transferred into the following invertible convolution relation:

$$\begin{aligned}\rho_c &= \int_{\Omega_I[\mathbf{f}]\times D_{\mathbf{v}}\times D_u} f(t,\mathbf{x},\mathbf{v},u)\,d\mathbf{x}\,d\mathbf{v}\,du\\ &=: \int_{\Omega_I[\mathbf{f};R]\times D_{\mathbf{v}}\times D_u} f(t,\mathbf{x},\mathbf{v},u)\,d\mathbf{x}\,d\mathbf{v}\,du.\end{aligned} \tag{3.48}$$

The domain Ω_I, which depends on time and $\mathbf{f}$, needs to be referred to the effective visibility zone Ω_S. In fact, the dynamics can be influenced by the relation between these two quantities, see [52]. As already mentioned in Subsection 3.3.5, if $\Omega_I \subseteq \Omega_S$, interactions occur only with a limited, generally fixed, number of particles that might

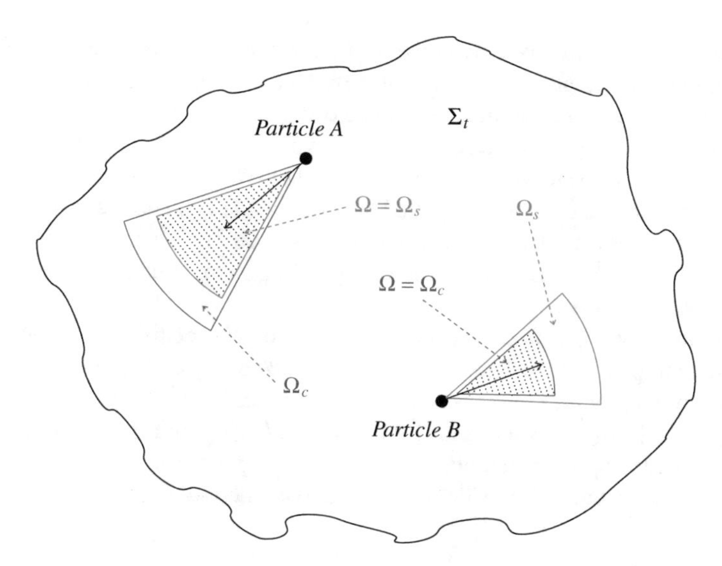

Fig. 3.5 Visibility and sensitivity zones for an arc of circle domain

be not all those within the sensitivity domain; however, the particle receives sufficient information to fully develop the standard strategy without restrictions. On the other hand, when $\Omega_S \subset \Omega_I$, interactions are not sufficient to fully develop their strategy. We adopt the notation $\Omega[\mathbf{f}, \mathbf{x}] = \Omega_I[\mathbf{f}] \bigcap \Omega_S(\mathbf{x})$ to denote the effective interaction domain [52].

In some special cases, this domain might be equal to zero so that particles do not modify their trajectory. We emphasize that the dependence on $\mathbf{f}$ makes the domain dynamic, not static. The role of the interaction domains is shown, related to the aforesaid concepts, in Fig. 3.5, where in cases (a) and (b) the needed information is missing, while it is not missing in case (c). This figure is somewhat "abstract" as the real shape of the interaction domains is related to the features of the specific system under consideration.

Distance between interacting particles: The following different concepts of distance can play an important role in the interaction dynamics, focusing specifically on the encounter rate:

(i) *State distance*, which simply depends on the distance

$$||(\mathbf{x}_*, \mathbf{v}_*, u_*) - (\mathbf{x}^*, \mathbf{v}^*, u^*)||$$

between the microscopic states, with a suitable metric to be considered for each specific case.

(ii) *Individual-mean state distance*, which is a different interpretation of the previous one and refers the state $\mathbf{x}_*, \mathbf{v}_*, u_*$ to the mean value of $\mathbf{x}_*, \mathbf{v}_*, u_*$ in the domain of interactions of the field particles according to a suitable metric. Such a distance can involve only some of the components of the microscopic state and can be formally denoted as follows: $D(\mathbf{x}_*, \mathbf{v}_*, u_*, \mathbb{E}_\Omega(\mathbf{x}_*, \mathbf{v}_*, u_*))$, where $\mathbb{E}_\Omega(\mathbf{x}_*, \mathbf{v}_*, u_*)$ denotes the mean value of $\mathbf{x}_*, \mathbf{v}_*, u_*$ in the domain Ω.

(iii) *Hierarchic distance*, which occurs when two active particles belong to different functional subsystems. Then, the distance $|k - h|$ can be defined if a conceivable numbering criterion is applied in selecting the first subsystem by a certain selection rule (for instance, in the animal world, the "dominant") and in numbering the others by increasing numbers depending on the decreasing rate. Namely, $|k - h| \uparrow \Rightarrow \eta_{hk} \downarrow$.

(iv) *Affinity distance*, when the distance between the distribution functions of interacting active particles can be defined according to the general idea that two systems with close distributions are *affine*. In this case, the distance is $||f_h - f_k||_{L^p(\Omega[\mathbf{f}])}$, where a uniform, e.g., $L_\infty(\Omega[\mathbf{f}])$, or mean, e.g., $L_1(\Omega[\mathbf{f}])$, approximation can be adopted as the most appropriate according to the physics of the system under consideration. As above, the rate decays with increasing distance.

Output of the interaction: An active particle interacts with all particles in its interaction domain Ω. The output of the interaction depends on the set of distribution functions $\mathbf{f}$ within Ω. For instance, it can depend on the moments computed for the particles within the interaction domain as shown in the modeling of swarms [52], where individuals feel both the stimuli of a fixed number of surrounding entities as well as that coming from the whole swarm (for instance, a clustering action toward the center of mass of the swarm).

In addition, the occurrence of a type of interaction with respect to the other, for instance cooperation or altruism, can be ruled, in several models, by a threshold on the distance between the interacting particles, see [60] for a model of opinion formation. On the other hand, it is shown in [46] that this type of dynamics depends on that of the overall system and hence on the set of the distribution functions $\mathbf{f}$.

Critical (small or large) size toward survival of functional subsystems: The survival of functional subsystems might, in some cases, depend on their size. If the size $n[f_i]$ of a functional subsystem falls below a critical value N_m, then interactions reduce to transfer individuals from the original one toward an aggregation to another one. In principles, a large critical size N_m might exist such that if the said threshold is overcome, particles are induced to move into another functional subsystem. This type of dynamics occurs in various social systems, where functional subsystems are identified by groups of interest [4]; for instance, in the case of voting dynamics, the critical size is a minimal number of entities that assure the survival of the party; when this size is not reached, supporters move from that party to an other.

Interaction rate: The interaction rate in the case of classical particles is simply given by the relative velocity of the interacting pairs, say $|\mathbf{v}_* - \mathbf{v}^*|$ or $|\mathbf{v} - \mathbf{v}^*|$ multiplied by some constants. However, active particles react to the other particles in their sensitivity zone according to an interaction rate that generally decays with the distance. A possible model is as follows:

$$\eta_{hk}[\mathbf{f}] = \eta^0 \exp\big\{ - d_{hk}[\mathbf{f}](\mathbf{x}_*, \mathbf{v}_*, u_*, \mathbf{x}^*, \mathbf{v}^*, u^*)\big\}, \tag{3.49}$$

with

$$d_{hk}[\mathbf{f}] = c_1|\mathbf{x}_*, \mathbf{v}_*, u_* - \mathbf{x}^*, \mathbf{v}^*, u^*| + c_2|h-k|$$

$$+c_3|\mathbf{x}_*, \mathbf{v}_*, u_* - \mathbb{E}_\Omega(\mathbf{x}^*, \mathbf{v}^*, u^*)| + c_4||f_h - f_k||_{L^p(\Omega[\mathbf{f}])}, \tag{3.50}$$

where linear interactions correspond to $c_3 = c_4 = 0$, while the same approach can be used, as we shall see later, for the interactions between the inner system and external agents with decay constants that can differ from that of η_{hk}. This, very simple, expression to estimate the distance in individual-based interactions can be definitely be improved by operating in Weierstrass spaces [233, 234].

Conservative and non-conservative interactions: The modeling of interactions in the case of particles heterogeneously distributed in space presents some important additional difficulties with respect to the spatially homogeneous case. In fact, each particle interacts with the particles in its interaction domain. The output of the interaction can be still related to the games described in Subsection 3.3.1, namely *consensus, competition, hiding, learning*, or others consistent with the specific system under consideration. In general, the output of the game can depend not only on the microstate of the interacting particles, but also on the set of distribution functions $\mathbf{f}$ within the interaction domain. As an example, it may depend on the moments computed for the particles within the interaction domain, say mean value or low-order moments.

Moreover, the same system, as already mentioned in Subsection 3.3.1, can undergo different types of interactions such as consensus against competition, where the occurrence of a type of interaction with respect to the other is ruled by a threshold on the distance between the interacting particles. The concept of distance used for the encounter rate can be used also in this case. For instance, if $d_{hk} \le \gamma$, competition occurs, while cooperation appears for $d_{hk} > \gamma$, where γ is a threshold related to the properties of the system. Some specific applications have shown [46] that this threshold depends on $\mathbf{f}$. Hence, the notation$\gamma[\mathbf{f}]$ is used. These reasoning can be referred both to conservative and to proliferative–destructive interactions.

Remark 3.13 *The reasonings concerning some of the above terms can be extended also to those modeling proliferation or destruction, where the encounter rate should differ for each type of specific dynamics. More in detail, the rate of proliferation or destruction depends on the affinity of the interacting pairs, namely on the distance in norm of their distribution functions. Analogous rationale can be applied to systems interacting with the outer environment. A more detailed modeling can be referred only for each specific case study as substantial differences can characterize different systems. The study of human crowd, presented in Chapter 6, will show some of the various possible modeling approaches in the case of number conservative systems.*

3.5.4 Mathematical structures

Bearing all above in mind, let us consider a large system of interacting active particles subdivided into m functional subsystems labeled by the subscript i. As already mentioned, a general mathematical structure suitable to describe the dynamics of the distribution functions f_i is obtained by a balance of the number of particles in the elementary space of the microscopic states. The dynamics, in the said microvolume and corresponding to the interactions described in Sect. 3.5.3, is the following:

Variation rate of the number of active particles
= *Inlet flux rate caused by conservative interactions*
−*Outlet flux rate caused by conservative interactions*
+*Inlet flux rate caused by proliferative interactions*
−*Outlet flux rate caused by destructive interactions*
−*natural decay toward an equilibrium solution*,

where the inlet flux includes the dynamics of mutations.

This flowchart, where the definition of the terms η, $\mathscr{A}$, $\mathscr{P}$, and $\mathscr{D}$ were given in Subsection 3.5.2, corresponds to the following structure:

$$\begin{aligned}\left(\partial_t + \mathbf{v}\cdot\nabla_{\mathbf{x}}\right) f_i(t,\mathbf{x},\mathbf{v},u) &= J_i[\mathbf{f}](t,\mathbf{x},\mathbf{v},u)\\ &= C_i[\mathbf{f}](t,\mathbf{x},\mathbf{v},u) + P_i[\mathbf{f}](t,\mathbf{x},\mathbf{v},u) - D_i[\mathbf{f}](t,\mathbf{x},\mathbf{v},u)\\ &- \lambda_i[f_i](f_i - f_{ie})(t,\mathbf{x},\mathbf{v},u),\end{aligned} \tag{3.51}$$

where a transport term has been included in the left-hand side of the above equation and is equated to the terms modeling conservative, proliferative, and destructive interactions and to the decay term.

The "gain" and "loss" terms corresponding to the conservative interactions, modeled by the term C_i, that modify the microstate, but not the number of particles, are given by the following:

$$\begin{aligned}C_i[\mathbf{f}](t,\mathbf{x},\mathbf{v},u) = \sum_{h,k=1}^{m}\int_{(\Omega\times D_u^2\times D_{\mathbf{v}}^2)[\mathbf{f}]} &\eta_{hk}[\mathbf{f}](\mathbf{x},\mathbf{x}^*,\mathbf{v}_*,\mathbf{v}^*,u_*,u^*)\\ &\times\mathscr{A}_{hk}^{i}[\mathbf{f}](\mathbf{v}_*\to\mathbf{v},\ u_*\to u|\mathbf{v}_*,\mathbf{v}^*,u_*,u^*)\\ &\times f_h(t,\mathbf{x},\mathbf{v}_*,u_*)f_k(t,\mathbf{x}^*,\mathbf{v}^*,u^*)\,d\mathbf{x}^*\,d\mathbf{v}_*\,d\mathbf{v}^*\,du_*\,du^*\end{aligned}$$

$$-f_i(t,\mathbf{x},\mathbf{v},u)\sum_{k=1}^{m}\int_{(\Omega\times D_u\times D_\mathbf{v})[\mathbf{f}]}\eta_{ik}[\mathbf{f}](\mathbf{x},\mathbf{x}^*,\mathbf{v},\mathbf{v}^*,u,u^*)$$

$$\times f_k(t,\mathbf{x}^*,\mathbf{v}^*,u^*)\,d\mathbf{x}^*\,d\mathbf{v}^*\,du^*. \tag{3.52}$$

The terms corresponding to proliferation and destruction dynamics are given by:

$$P_i[\mathbf{f}](t,\mathbf{x},\mathbf{v},u)=\sum_{h,k=1}^{m}\int_{(\Omega\times D_u^2\times D_\mathbf{v}^2)[\mathbf{f}]}\eta_{hk}[\mathbf{f}](\mathbf{x},\mathbf{x}^*,\mathbf{v}_*,\mathbf{v}^*,u_*,u^*)\,\mathscr{P}_{hk}^{i}[\mathbf{f}](u_*\to u|u^*,u_*)$$

$$\times f_h(t,\mathbf{x},\mathbf{v}_*,u_*)f_k(t,\mathbf{x}^*,\mathbf{v}^*,u^*)\,d\mathbf{x}^*\,d\mathbf{v}_*\,d\mathbf{v}^*\,du_*\,du^*, \tag{3.53}$$

and

$$D_i[\mathbf{f}](t,\mathbf{x},\mathbf{v},u)=f_i(t,\mathbf{x},\mathbf{v},u)\sum_{k=1}^{m}\int_{(\Omega\times D_u\times D_\mathbf{v})[\mathbf{f}]}\eta_{ik}[\mathbf{f}](\mathbf{x},\mathbf{x}^*,\mathbf{v},\mathbf{v}^*,u,u^*)$$

$$\times\,\mathscr{D}_{ik}[\mathbf{f}](u,u^*)\,f_k(t,\mathbf{x}^*,\mathbf{v}^*,u^*)\,d\mathbf{x}^*\,d\mathbf{v}^*\,du^*, \tag{3.54}$$

under the assumption, $\mathscr{P}_{hk}^{i}$ and $\mathscr{D}_{ik}$ depend on the activity variables only.

Substituting these terms into Eq. 3.51 yields the derivation of a general mathematical structure.

Reasoning analogous to those developed in Sect. 3.3 leads to a mathematical structure for systems subject to an external actions. Suppose that external actions are applied, at the macroscopic level, by a field $\mathbf{F}_i(t,\mathbf{x})$ acting over the mechanical variables and by $\mathbf{K}_i(t,u)$ acting over the activity variable, the corresponding structure is as follows:

$$\big(\partial_t+\mathbf{v}\cdot\nabla_\mathbf{x}+\mathbf{F}_i(t,\mathbf{x})\cdot\nabla_\mathbf{v}+\partial_u\mathbf{K}_i(t,u)\big)\,f_i(t,\mathbf{x},\mathbf{v},u)=J_i[\mathbf{f}](t,\mathbf{x},\mathbf{v},u)$$

$$=C_i[\mathbf{f}](t,\mathbf{x},\mathbf{v},u)+P_i[\mathbf{f}](t,\mathbf{x},\mathbf{v},u)-D_i[\mathbf{f}](t,\mathbf{x},\mathbf{v},u)$$

$$-\,\lambda_i[f_i](f_i-f_{ie})(t,\mathbf{x},\mathbf{v},u). \tag{3.55}$$

If the external action is applied by agents with a prescribed state, which may depend on time, space, and activity, $g_i=g_i(t,\mathbf{x},\mathbf{v},u)$, then the structure takes the following form:

$$\big(\partial_t+\mathbf{v}\cdot\nabla_\mathbf{x}\big)\,f_i(t,\mathbf{x},\mathbf{v},u)=J_i[\mathbf{f}](t,\mathbf{x},\mathbf{v},u)+I_i[\mathbf{f}](t,\mathbf{x},\mathbf{v},u), \tag{3.56}$$

where J_i has been defined by Eqs. (3.51)–(3.54), while I_i can be computed according to the same approach. The result, with obvious meaning of notations, is as follows:

$$I_i[\mathbf{f}](t, \mathbf{x}, \mathbf{v}, u) = C_i^e[\mathbf{f}](t, \mathbf{x}, \mathbf{v}, u) + P_i^e[\mathbf{f}](t, \mathbf{x}, \mathbf{v}, u) - D_i^e[\mathbf{f}](t, \mathbf{x}, \mathbf{v}, u), \quad (3.57)$$

where

$$\begin{aligned} C_i^e[\mathbf{f}](t, \mathbf{x}, \mathbf{v}, u) = \sum_{h,k=1}^{m} \int_{(\Omega \times D_u^2 \times D_\mathbf{v}^2)[\mathbf{f}]} & \mu_{hk}[\mathbf{f}](\mathbf{x}, \mathbf{x}^*, \mathbf{v}_*, \mathbf{v}^*, u_*, u^*) \\ & \times \mathscr{B}_{hk}^i[\mathbf{f}](\mathbf{v}_* \to \mathbf{v},\ u_* \to u | \mathbf{x}, \mathbf{x}^*, \mathbf{v}_*, \mathbf{v}^*, u_*, u^*) \\ & \times f_h(t, \mathbf{x}, \mathbf{v}_*, u_*) g_k(t, \mathbf{x}^*, \mathbf{v}^*, u^*)\, d\mathbf{x}^*\, d\mathbf{v}_*\, d\mathbf{v}^*\, du_*\, du^* \\ & - f_i(t, \mathbf{x}, \mathbf{v}, u) \sum_{k=1}^{m} \int_{(\Omega \times D_u \times D_\mathbf{v})[\mathbf{f}]} \mu_{ik}[\mathbf{f}](\mathbf{x}, \mathbf{x}^*, \mathbf{v}, \mathbf{v}^*, u, u^*) \\ & \times g_k(t, \mathbf{x}^*, \mathbf{v}^*, u^*)\, d\mathbf{x}^*\, d\mathbf{v}^*\, du^*, \end{aligned} \quad (3.58)$$

$$\begin{aligned} P_i^e[\mathbf{f}](t, \mathbf{x}, \mathbf{v}, u) = \sum_{h,k=1}^{m} \int_{(\Omega \times D_u^2 \times D_\mathbf{v}^2)[\mathbf{f}]} & \mu_{hk}[\mathbf{f}](\mathbf{x}, \mathbf{x}^*, \mathbf{v}_*, \mathbf{v}^*, u_*, u^*) \\ & \times \mathscr{P}_{hk}^{i,e}[\mathbf{f}](u_* \to u | u_*, u^*) \\ & \times f_h(t, \mathbf{x}, \mathbf{v}_*, u_*) g_k(t, \mathbf{x}^*, \mathbf{v}^*, u^*)\, d\mathbf{x}^*\, d\mathbf{v}_*\, d\mathbf{v}^*\, du_*\, du^*, \end{aligned} \quad (3.59)$$

and

$$\begin{aligned} D_i^e[\mathbf{f}](t, \mathbf{x}, \mathbf{v}, u) = f_i(t, \mathbf{x}, \mathbf{v}, u) \sum_{k=1}^{m} \int_{(\Omega \times D_u \times D_\mathbf{v})[\mathbf{f}]} & \mu_{ik}[\mathbf{f}](\mathbf{x}, \mathbf{x}^*, \mathbf{v}, \mathbf{v}^*, u, u^*) \\ & \times \mathscr{D}_{ik}^e[\mathbf{f}](u, u^*)\, g_k(t, \mathbf{x}^*, \mathbf{v}^*, u^*)\, d\mathbf{x}^*\, d\mathbf{v}^*\, du^*, \end{aligned} \quad (3.60)$$

with

- $\mu_{hk}[\mathbf{f}](\mathbf{x}, \mathbf{v}_*, u_*, \mathbf{x}^*, \mathbf{v}^*, u^*)$ models the frequency of the interactions between a candidate h-particle with state $\mathbf{x}, \mathbf{v}_*, u_*$ and a field k-agent with state $\mathbf{x}^*, \mathbf{v}^*, u^*$.
- $\mathscr{B}_{hk}^i[\mathbf{f}](\mathbf{x}, \mathbf{v}_*, u_* \to \mathbf{x}, \mathbf{v}, u | \mathbf{x}, \mathbf{x}^*, \mathbf{v}, \mathbf{v}_*, u, u_*)$ denotes the probability density that a candidate h-particle, with state $\mathbf{x}, \mathbf{v}_*, u_*$, ends up into the state of the test particle of the ith functional subsystem after an interaction with a field k-agent.

- $P_{hk}^{i,e}[\mathbf{f}](u_* \to u|u_*, u^*)$ models the proliferative events for a candidate h-particle, with state $\mathbf{x}, \mathbf{v}_*, u_*$, into the ith functional subsystem after interaction with a field k-agent with state $\mathbf{x}^*, \mathbf{v}^*, u^*$.
- *Destructive term* $\mathscr{D}_{ik}^{e}[\mathbf{f}](u, u^*)$ models the rate of destruction for a test i-particle in its own functional subsystem after an interaction with a field k-agent with state $\mathbf{x}, \mathbf{v}^*, u^*$.

The mathematical structures (3.51)–(3.60) offer a formal framework, while a detailed modeling of the interaction terms is needed to obtain specific models.

A practical example is given by models which include conservative interactions only:

$$
\begin{aligned}
&\left(\partial_t + \mathbf{v} \cdot \nabla_{\mathbf{x}}\right) f_i(t, \mathbf{x}, \mathbf{v}, u) \\
&\quad = \sum_{h,k=1}^{m} \int_{(\Omega \times D_u^2 \times D_{\mathbf{v}}^2)[\mathbf{f}]} \eta_{hk}[\mathbf{f}](\mathbf{x}, \mathbf{x}^*, \mathbf{v}_*, \mathbf{v}^*, u_*, u^*) \\
&\qquad \times \mathscr{A}_{hk}^{i}[\mathbf{f}](\mathbf{v}_* \to \mathbf{v},\ u_* \to u|\mathbf{x}, \mathbf{x}^*, \mathbf{v}_*, \mathbf{v}^*, u_*, u^*) \\
&\qquad \times f_h(t, \mathbf{x}, \mathbf{v}_*, u_*) f_k(t, \mathbf{x}^*, \mathbf{v}^*, u^*)\, d\mathbf{x}^*\, d\mathbf{v}_*\, d\mathbf{v}^*\, du_*\, du^* \\
&\quad - f_i(t, \mathbf{x}, \mathbf{v}, u) \sum_{k=1}^{m} \int_{(\Omega \times D_u \times D_{\mathbf{v}})[\mathbf{f}]} \eta_{ik}[\mathbf{f}](\mathbf{x}, \mathbf{x}^*, \mathbf{v}, \mathbf{v}^*, u, u^*) \\
&\qquad \times f_k(t, \mathbf{x}^*, \mathbf{v}^*, u^*)\, d\mathbf{x}^*\, d\mathbf{v}^*\, du^* \\
&\quad + \sum_{h,k=1}^{m} \int_{(\Omega \times D_u^2 \times D_{\mathbf{v}}^2)[\mathbf{f}]} \mu_{hk}[\mathbf{f}](\mathbf{x}, \mathbf{x}^*, \mathbf{v}_*, \mathbf{v}^*, u_*, u^*) \\
&\qquad \times \mathscr{B}_{hk}^{i}[\mathbf{f}](\mathbf{v}_* \to \mathbf{v},\ u_* \to u|\mathbf{v}_*, \mathbf{v}^*, u_*, u^*) \\
&\qquad \times f_h(t, \mathbf{x}, \mathbf{v}_*, u_*) g_k(t, \mathbf{x}^*, \mathbf{v}^*, u^*)\, d\mathbf{x}^*\, d\mathbf{v}_*\, d\mathbf{v}^*\, du_*\, du^* \\
&\quad - f_i(t, \mathbf{x}, \mathbf{v}, u) \sum_{k=1}^{m} \int_{(\Omega \times D_u \times D_{\mathbf{v}})[\mathbf{f}]} \mu_{ik}[\mathbf{f}](\mathbf{x}, \mathbf{x}^*, \mathbf{v}, \mathbf{v}^*, u, u^*) \\
&\qquad \times g_k(t, \mathbf{x}^*, \mathbf{v}^*, u^*)\, d\mathbf{x}^*\, d\mathbf{v}^*\, du^* \\
&\qquad - \lambda_i[f_i](f_i - f_{ie})(t, \mathbf{x}, \mathbf{v}, u).
\end{aligned}
\tag{3.61}
$$

Similarly, models which include proliferative/destructive dynamics only show the following structure:

$$
\begin{aligned}
&\left(\partial_t + \mathbf{v}\cdot\nabla_{\mathbf{x}}\right) f_i(t,\mathbf{x},\mathbf{v},u) \\
&\quad = \sum_{h,k=1}^{m} \int_{(\Omega\times D_u^2\times D_{\mathbf{v}}^2)[\mathbf{f}]} \eta_{hk}[\mathbf{f}](\mathbf{x},\mathbf{x}^*,\mathbf{v}_*,\mathbf{v}^*,u_*,u^*)\, \mathscr{P}^i_{hk}[\mathbf{f}](u_*\to u|u_*,u^*) \\
&\qquad \times f_h(t,\mathbf{x},\mathbf{v}_*,u_*) f_k(t,\mathbf{x}^*,\mathbf{v}^*,u^*)\, d\mathbf{x}^*\, d\mathbf{v}_*\, d\mathbf{v}^*\, du_*\, du^* \\
&\quad - f_i(t,\mathbf{x},\mathbf{v},u) \sum_{k=1}^{m} \int_{(\Omega\times D_u\times D_{\mathbf{v}})[\mathbf{f}]} \eta_{ik}[\mathbf{f}](\mathbf{x},\mathbf{x}^*,\mathbf{v},\mathbf{v}^*,u,u^*) \\
&\qquad \times \mathscr{D}_{ik}[\mathbf{f}](u,u^*)\, f_k(t,\mathbf{x}^*,\mathbf{v}^*,u^*)\, d\mathbf{x}^*\, d\mathbf{v}^*\, du^* \\
&\quad + \sum_{h,k=1}^{m} \int_{(\Omega\times D_u^2\times D_{\mathbf{v}}^2)[\mathbf{f}]} \mu_{hk}[\mathbf{f}](\mathbf{x},\mathbf{x}^*,\mathbf{v}_*,\mathbf{v}^*,u_*,u^*)\, \mathscr{P}^{i,e}_{hk}[\mathbf{f}](u_*\to u|u_*,u^*) \\
&\qquad \times f_h(t,\mathbf{x},\mathbf{v}_*,u_*) g_k(t,\mathbf{x}^*,\mathbf{v}^*,u^*)\, d\mathbf{x}^*\, d\mathbf{v}_*\, d\mathbf{v}^*\, du_*\, du^* \\
&\quad - f_i(t,\mathbf{x},\mathbf{v},u) \sum_{k=1}^{m} \int_{(\Omega\times D_u\times D_{\mathbf{v}})[\mathbf{f}]} \mu_{ik}[\mathbf{f}](\mathbf{x},\mathbf{x}^*,\mathbf{v},\mathbf{v}^*,u,u^*) \\
&\qquad \times \mathscr{D}^e_{ik}[\mathbf{f}](u,u^*)\, f_k(t,\mathbf{x}^*,\mathbf{v}^*,u^*)\, d\mathbf{x}^*\, d\mathbf{v}^*\, du^* \\
&\qquad - \lambda_i[f_i](f_i - f_{ie})(t,\mathbf{x},\mathbf{v},u).
\end{aligned} \tag{3.62}
$$

Models with space structure have been applied in various fields such as vehicular traffic and crowd dynamics. Details and bibliography are given in Chapter 6 but notice should be made that the literature is limited to the case of systems in absence of external actions. Therefore, the framework presented in this subsection has to be considered as a formal proposal to be properly developed and adjusted to specific applications.

3.5.5 *Mean field models*

Although this book focuses on mathematical structures with interactions modeled by theoretical tools of game theory, it is worth outlining a conceivable alternative based on continuous field actions over active particles. Some guidelines are given without tackling specific modeling problems. This approach requires the modeling of the microscale actions

$$h_i[\mathbf{f}](\mathbf{x}, \mathbf{v}, u, \mathbf{x}^*, \mathbf{v}^*, u^*) \quad \text{and} \quad k_i[\mathbf{f}](\mathbf{x}, \mathbf{v}, u, \mathbf{x}^*, \mathbf{v}^*, u^*)$$

acting on the velocity and activity variables, respectively, of the field particles with state $\mathbf{x}^*, \mathbf{v}^*, u^*$ which are in the domain $(\Omega \times D_{\mathbf{v}} \times D_u)[\mathbf{f}]$ and of the test particle with state $\mathbf{x}, \mathbf{v}, u$. These actions depend, in the general case, on the microstates and distribution functions of the interacting pairs.

The overall action is obtained by integration over the interaction domain:

$$H_i[\mathbf{f}](t, \mathbf{x}, \mathbf{v}, u) = \int_{(\Omega \times D_{\mathbf{v}} \times D_u)[\mathbf{f}]} h_i[\mathbf{f}](\mathbf{x}, \mathbf{v}, u, \mathbf{x}^*, \mathbf{v}^*, u^*)\, d\mathbf{x}^*\, d\mathbf{v}^*\, du^*, \tag{3.63}$$

and

$$K_i[\mathbf{f}](t, \mathbf{x}, \mathbf{v}, u) = \int_{(\Omega \times D_{\mathbf{v}} \times D_u)[\mathbf{f}]} k_i[\mathbf{f}](\mathbf{x}, \mathbf{v}, u, \mathbf{x}^*, \mathbf{v}^*, u^*)\, d\mathbf{x}^*\, d\mathbf{v}^*\, du^*. \tag{3.64}$$

When this project can be developed, a Vlasov-type equation can be written as it is documented in a rapidly growing literature on swarming phenomena [77, 175, 180] generated from some pioneer papers in the field [191]. A general expression is as follows:

$$\partial_t f_i + \mathbf{v} \cdot \partial_{\mathbf{x}} f_i + \partial_{\mathbf{v}}(H_i[\mathbf{f}] f_i) + \partial_u (K_i[\mathbf{f}] f_i) = 0, \tag{3.65}$$

where $f_i = f_i(t, \mathbf{x}, \mathbf{v}, u)$.

3.5.6 *Perturbation of space homogeneity*

An additional approach to derive a differential structure, where the space variable is included, consists in modeling a stochastic velocity perturbation of the spatially homogeneous dynamics. This approach was introduced in [193] to derive macroscopic transport equations from the underlying description delivered by kinetic theory models. This approach has been further developed by various authors mainly looking at applications in biology. The related, already vast bibliography is reported in Section 4 of the survey [33]. The method seems appropriate for biological tissues, while it does not yet appear useful in the modeling of crowd dynamics.

Let us now briefly report the derivation of the aforementioned mathematical structure referring to a system of interacting functional subsystems, whose state is encoded in the *distribution function*. The mathematical structure can be obtained by adding to the space homogeneous dynamics, a stochastic perturbation in velocity, which induces transport

$$(\partial_t + \mathbf{v} \cdot \nabla_{\mathbf{x}})\; f_i = \nu_i\, L_i(f_i) + J_i[\mathbf{f}], \tag{3.66}$$

where the perturbation operators $L_i[f_i]$ are given by

$$L_i(f_i) = \int_V \Big[T_i(\mathbf{v}, \mathbf{v}^*) f_i(t, \mathbf{x}, \mathbf{v}^*, u) - T_i(\mathbf{v}^*, \mathbf{v}) f_i(t, \mathbf{x}, \mathbf{v}, u) \Big] \, d\mathbf{v}^*, \tag{3.67}$$

and where $T_i(\mathbf{v}, \mathbf{v}^*)$ is the transition probability kernel over the new velocity $\mathbf{v} \in V$, assuming that the previous velocity is $\mathbf{v}^*$. This corresponds to the assumption that any individual of the population chooses any direction with bounded velocity. The set of possible velocities is denoted by V, where $V \subset I\!R^3$; moreover, it is assumed that V is bounded and spherically symmetric (i.e., $v \in V \Rightarrow -v \in V$); ν_i is the corresponding turning rate.

3.6 Critical Analysis

The mathematical structures derived in Sections 3.4 and 3.5 offer a formal framework deemed to capture the complexity features of living systems. This is the first step of the modeling approach, while the second step should develop a detailed modeling of interactions that is needed to obtain specific models by implementing the said interaction models into the general mathematical structure.

Indeed, these structures play the role of field theories that cannot be used in the case of living systems. The conceptual differences with respect to the classical kinetic theory have been enlightened all along the preceding sections. Therefore, it is now useful examining how far the said structures retain the complexity features of living systems presented in Section 1.4.

1. *Large number of components* : The functional system approach selects those parts of the overall system, which play effectively the role. Therefore, the selection of the dependent variables related to each functional subsystem accounts also for the need to reduce complexity induced by the generally *large number of components*. This avoids dealing with a number of equations necessary that might be too large to be practically treated.

2. *Ability to express a strategy* : This ability is related to the activity variable and to the games which model conservative interactions generally based on rational behaviors. However, in special conditions, for instance panic, also irrational ones even by a few entities can induce large deviations with respect to the usual dynamics corresponding to rational behaving. The strategy can also be expressed by proliferative and/or destructive interactions which generate new particles or model their destruction. Generation of particles might even occur in functional subsystems different from that of the generating particles. Therefore, the number of equations might evolve in time.

3. *Heterogeneity* : The said ability is *heterogeneously distributed* as the dependent variables, deemed to describe the overall state of the system, are probability distributions. More precisely, when the number of individuals is constant in time, it can be a probability density, while when proliferative and/or destructive encounters appear, it is a distribution function. This feature is taken into account by the activity variable and by the description of the system by probability distributions rather than by deterministic variables.

4. *Behavioral stochastic rules* : *Stochastic features* are not limited to the selection of the dependent variables, but are included in the modeling of interactions, where players are probability distributions and the output of interactions is modeled by stochastic rules related to individual rational and/or irrational behaviors. These interactions can be classified as doubly stochastic as the state of the interacting pairs is defined by a probability distribution and the output is also modeled by stochastic rules.

5. *Nonlinear interactions* : The mathematical structure includes *nonlinearly additive and non-local interactions*, namely the output of interactions can depend also on the whole probability distribution rather than on the microscopic and independent variables. Nonlinearity refers also to the interaction rate as, in some cases, the topological distribution of a fixed number of neighbors can play a prominent role in the development of strategy and interactions. Namely, living entities interact, in certain physical conditions, with a fixed number of entities rather than all those in their visibility domain. Non-locality is an important feature of interaction models.

6. *Learning dynamics* : The modeling needs taking into account the *ability of living systems to learn from past experience*. A general theory is not yet available, while the applications presented in the next chapters will give an account of the state of the art. However, the rules of the games modeling interactions have to be allowed to evolve in time also due to learning ability. It is worth stressing that learning dynamics can involve the whole population and not only individuals.

7. *Darwinian mutations and selection* : Mathematical tools include the possibility of describing *Darwinian mutations and selection* in each birth process generation, and mutations bring new genetic variants into populations. Natural selection then screens them: The rigors of the environment reduce the frequency of "bad" (not well fitted with respect to the others) variants and increase the frequency of "good" (well fitted with respect to the others) ones. The analogy with social systems deserves to be deeply studied, an account of this feature is given in Chapter 4, and therefore, post-Darwinian theories draw a parallel between sociology and developmental biology.

8. *Multiscale features* : As stated already in Chapter 1, the modeling approach always needs *multiscale methods*. The mathematical structures proposed in this chapter link the dynamics at the individual based, namely microscopic, scale to the overall dynamics. However, additional work still needs to be developed in order to treat exhaustively this topic. As an example, some preliminary results are known

mainly in biology, where the dynamics at the molecular (genetic) level determines the ability of cells to express specific functions. In addition, the structure of macroscopic tissues depends on the dynamics at the cellular scale, and hence, their derivation needs to be developed by appropriate asymptotic methods. The next chapters will present some bibliography on this challenging mathematical problem.

9. *Time is a key variable* : The validation of models should also take into account that there exists a timescale of observation and modeling of living systems which is long enough to observe evolutionary events that generally change substantially the strategy expressed by individuals. In general, the mathematical approach should set this observation and modeling lapse of time related to the specific objective of the investigation, and hence, mathematical models should provide a description of the system within this observation time. It is plain that Darwinian dynamics provides important indications on the setting of the observation and modeling lapse of time.

10. *Emerging behaviors* : Living systems show *collective emerging behaviors*, generally reproduced at a qualitative level given certain input conditions, though quantitative matches are rarely obtained by observations. Therefore, it is appropriate discussing them later when specific models will be presented. However, it is worth stressing that validation of models is based on not only their ability to depict empirical data obtained in steady cases, but also emerging behaviors that appear far from steady and equilibrium conditions. Some applications also refer to rare events, namely the so-called *Black Swan*.

The mathematical structures derived in this chapter can be regarded, as mentioned above, as a formal container of conceivable models. Sometimes, the term *mathematical theory* has been used in the literature considering that a new class of equations is proposed and that it substantially differs from that of the classical kinetic theory reviewed in Chapter 2. Most of the differences have been introduced to account the specific features of the living systems as critically analyzed in the preceding section. In particular, interactions are not reversible and follow stochastic rules within theoretical framework offered by evolutive game theory.

It is worth stressing that the introduction of the activity variable is not simply an artificial strategy that occurs in the classical kinetic theory when additional variables, e.g., a spin, are inserted to account further properties of the molecules. Here, it is part of the modeling approach as it is not only simply related to the strategy that living entities are able to express, but also to the partition of the overall system into functional subsystems.

However, it can be argued, at this end, that the reader has already received sufficient information to identify the substantial differences of the kinetic theory of classical particles from that of active particles. This analysis can be completed by understanding if the approach for active particles can include, as a special case, the classical theory. A reference on this topic is given by paper [12] for a class of equations simpler than those treated in this chapter. Arguably, the analysis can be extended to more general cases.

The concept of mathematical theory has been introduced in this section. However, it needs to be mentioned that one has to look ahead to further development of the said theory, for instance focusing on learning dynamics and multiscale topics. Some of these developments will be treated in the applications proposed in the next chapters, but are not yet organized within a general framework. Hence, one can look at them as a challenging research perspective. A preliminary step consists in looking at the existing approaches to complex systems and their interplay with the contents of this chapter.

For instance, *population dynamics* studies how the number of individuals in one or several interacting populations changes under the action of biological and environmental processes. The point of view is super-macroscopic: The elementary entities are the populations themselves as a whole, whose rise and decline over time are investigated. Models are classically stated in terms of systems of ordinary differential equations, whose unknowns are sizes of the various populations. Heterogeneity can be introduced by selecting populations featured by different behaviors. A substantial improvement is obtained by models of *population with internal structure* well settled in the mathematical framework by Webb [240], formalized through systems of partial differential equations. Recent applications are presented in the book by Perthame [199]. This approach introduces an additional variable, besides the number of individuals, describing an inner characteristics of the population, which is supposed to play a role in the emergence of collective behaviors (for instance, the age of the individuals, their fitness for the outer environment, and their social status).

As mentioned already, an important contribution to understand the dynamics of the systems under consideration is given by the theory of *evolutive games* [186], see also [214, 215] and the bibliography cited therein. These new theories study how players' strategy evolves in time due to selective processes, which can lead to clustering of the players into different groups depending on their fitness for the outer environment. The time evolution of game dynamics can be modeled in terms of differential games [69–71], where players apply a control action over basic dynamics modeled by ordinary differential equations in order to increase their payoffs.

Methods of statistical mechanics have been developed by Helbing, who understood that individual interactions need to be modeled by methods of game theory [138]. A unified approach, suitable to link methods of statistical mechanics, kinetic theory, and game theory, is proposed in the book [140]. A general purpose aims at modeling the overall dynamics of modern societies, where hard sciences such as mathematics and physics can contribute to improve the quality of life [141, 146].

A parallel, conceptually different, approach is that of the *kinetic theory presented in* [197], *which uses Boltzmann-type equations where the velocity is replaced by internal variables related to the specific social systems under consideration. A variety of interesting applications are presented in [197]*, while computations are developed by Monte Carlo particle methods as in our book.

Finally, let us stress that the mathematical structures presented in this chapter can be object of possible refinements and improvements whenever motivated by applications. The aim consists in supporting the research activity of applied mathematicians involved in modeling and applications of active particles methods. The various chapters of the recently published book [37] offer a broad variety of possible applications and research perspectives.

Chapter 4
From the Mathematical Theory to Applications

4.1 Plan of the Chapter

This chapter proposes a bridge to link the theoretical tools proposed in Chapter 3 to the various applications presented in the next chapters. In addition, some reasonings on the reply to the third and fourth key questions are brought to the attention of the reader. These can be viewed as a preliminary answer to be made practical and complete in the next chapters focusing on specific applications.

The flowchart of Figure 4.1 shows how two different classes of models will be studied in the next two chapters. In more detail, Chapter 5 focuses on models, where the microscopic state is limited to the activity variable only, while Chapter 6 proposes models, where the microscopic state includes both mechanical and activity variables; namely, interactions occur heterogeneously in space. This technical difference induces, as we shall see, important conceptual differences not only on the derivation of models, but also on their computational problems and validation.

The modeling approach is developed, first, by adapting the general mathematical structures proposed in Chapter 3 to the specific case under consideration and, subsequently, by deriving models after a detailed modeling of interactions. Once specific models have been derived, their application can generate analytic and computational problems, as well as to further revisions and enrichment of the aforementioned mathematical structures.

The structures proposed in Sections 3.3 and 3.4 correspond to space homogeneity and networks, as in the application of Chapter 5, while the structures of Section 3.5 correspond to the case of heterogeneous space behaviors as in the application of Chapter 6.

The flowchart indicates explicitly the focus of Chapter 5 on social systems and of Chapter 6 on crowd dynamics. In addition, each chapter will introduce the reader some further case studies of interest for the applications. The selection of the contents of the next two chapters and their common presentation style are consistent with the aim of the second part of the book devoted to address the future research activity in the specific fields of the applications.

N. Bellomo et al., *A Quest Towards a Mathematical Theory of Living Systems*, Modeling and Simulation in Science, Engineering and Technology,
DOI 10.1007/978-3-319-57436-3_4

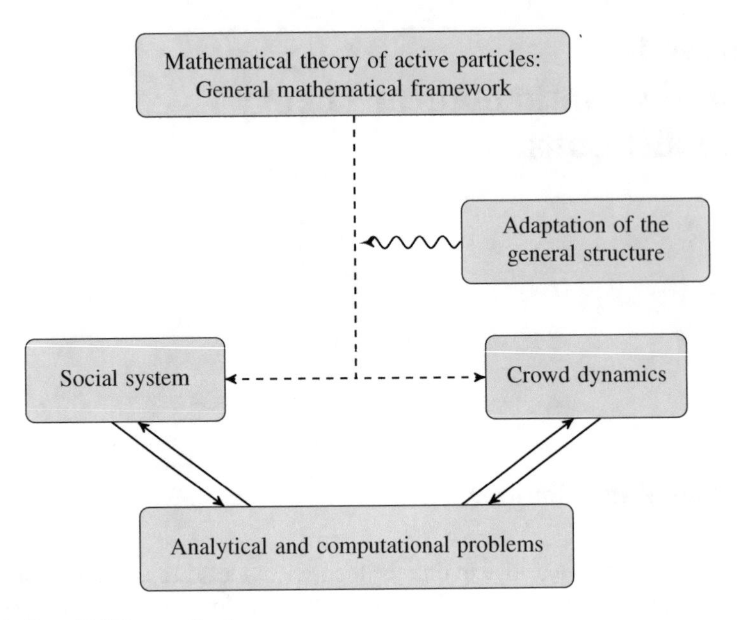

Fig. 4.1 From KTAP to applications

Due to this aim, the presentation will not be a technical revisiting of what is already known in the literature, but a critical analysis of what has been done is followed by constructive ideas on what should be done in a future research activity.

In addition, this chapter also poses the basis for the contents of Chapter 7, where it is shown how possible research programs, focused on modeling topics, can be developed.

In more detail, the contents of this chapter are as follows:

- Section 4.2 motivates the selection of the case studies, where the selection looks for applications characterized by different features and conceptual difficulties. The applications refer to social dynamics over networks and to the dynamics over a human crowd in complex venues.
- Section 4.3 provides some reasonings toward the answer to the third key question:

How mathematical models
can be referred to the mathematical structures
deemed to depict complexity features of living systems?

Then, this section shows the methodological approach by which models can be derived.

- Section 4.4 presents a preliminary rationale on the validation of models focusing precisely on the fourth key question:

Models offer a predictive ability,

$$but\ how\ can\ they\ be\ validated?$$
$$In\ addition,\ can\ rare\ events,$$
$$namely\ the\ so-called\ BlackSwans,\ be\ predicted\ by\ them?$$

- Section 4.5 develops a critical analysis which looks at further conceptual tasks related to modeling topics and applications.

4.2 Selection of the Case Studies

The selection of the applications treated in the next chapters has been planned with the aim of presenting applications featured by different conceptual difficulties to be tackled in the modeling approach. We will enlighten these differences also in view of the answer to the third and fourth key questions.

The first substantial difference, which distinguishes Chapter 5 from 6, is the way by which the role of the space variable is modeled. In the first case, space simply defines the localization of the nodes of the network, while the dynamics in each node is assumed to depend on time only. In the second case, the dynamics depends on space, in one or more dimensions, which is treated as a continuous variable.

Additional important differences can be pointed out focusing on interactions which are number preserving in the case of social dynamics extensively treated in Chapter 5, while proliferative and destructive dynamics have to be included in biological applications referring specifically to the immune competition.

The case study proposed in Chapter 5 is the development of criminality contrasted by security forces within the general framework of social dynamics. The specific system is number conservative; however, the immune competition is dealt with later in the chapter. In both cases, a social dynamics, which modifies the activity variable and a Darwinian type dynamics with mutation and selection, can be taken into account with different levels of intensity.

The case study proposed in Chapter 6 is the study of human crowds, where the presence of the space variable involves an additional number of technical difficulties. An additional feature that distinguishes this application is the role of the activity variable which can evolve in time and diffuse in space depending on the overall dynamics of the system under consideration.

The most known, and studied, case is the onset and propagation of panic. Indeed, the presence of this emotional state modifies the psychological attitude of walkers and modifies also the interaction rules. As a consequence, the collective dynamics can be substantially modified. This topic is further developed to propose some reasonings on the modeling of swarms and to show the conceptual differences in the modeling approach with respect to crowd dynamics.

Once the specific case studies have been selected, the presentation of both Chapters 5 and 6 is proposed along the same style according to the following index:

- Interpretation of the phenomenology of the class of systems object of the modeling approach focusing on its complexity features to understand how these features can influence the dynamics.
- An overview and critical analysis of what the existing literature offers including also specific applications.
- Revisiting the general structures proposed in Chapter 3 by adjusting it to the specific contents of each chapter. This specialization aims at offering the conceptual framework toward the derivation of specific models.
- Analysis of specific models and related simulations with special focus on enlightening the predictive ability of the models.
- Presentation of conceivable developments toward a modeling approach to be organized within appropriate research programs. Research perspectives refer also to analytic problems generated by applications.
- Generalizations of the modeling approach to a variety of applications.

4.3 From Mathematical Structures to the Derivation of Models

This section shows how mathematical models of living systems can be derived referring to the general mathematical structures derived in Chapter 3. Namely, how the said structures can be specialized according to a phenomenological interpretation of the specific systems under consideration.

The flowchart in Figure 4.2 shows the organization of the overall rationale and indicates how analytic and computational problems can contribute to further development of the mathematical theory, while details of the application of the modeling approach are as follows:

1. **Phenomenological observation and interpretation of the system:** *This analysis corresponds to the assessment of the functional subsystems which play the game and of their representation, namely the activity variable of each subsystem and the network where they operate.*
2. **Specialization of the mathematical structure:** *The phenomenological interpretation contributes to specialize the general mathematical structure to that needed by the modeling of each class of systems. This task is achieved by keeping only the interactions relevant to the dynamics of each system.*
3. **Modeling interactions:** *Interactions related to the structure defined in Step 2 are modeled by theoretical tools of evolutionary stochastic game theory such as those reported in Chapter* 3.
4. **Derivation of models:** *Implementing these models of interactions into the structure defined in Step 2 yields the mathematical model.*
5. **Parameter sensitivity analysis:** *Each model is characterized by parameters generally related to each specific type of interactions. A parameter sensitivity analysis is necessary to identify the parameters which have an effective influence on the overall dynamics.*

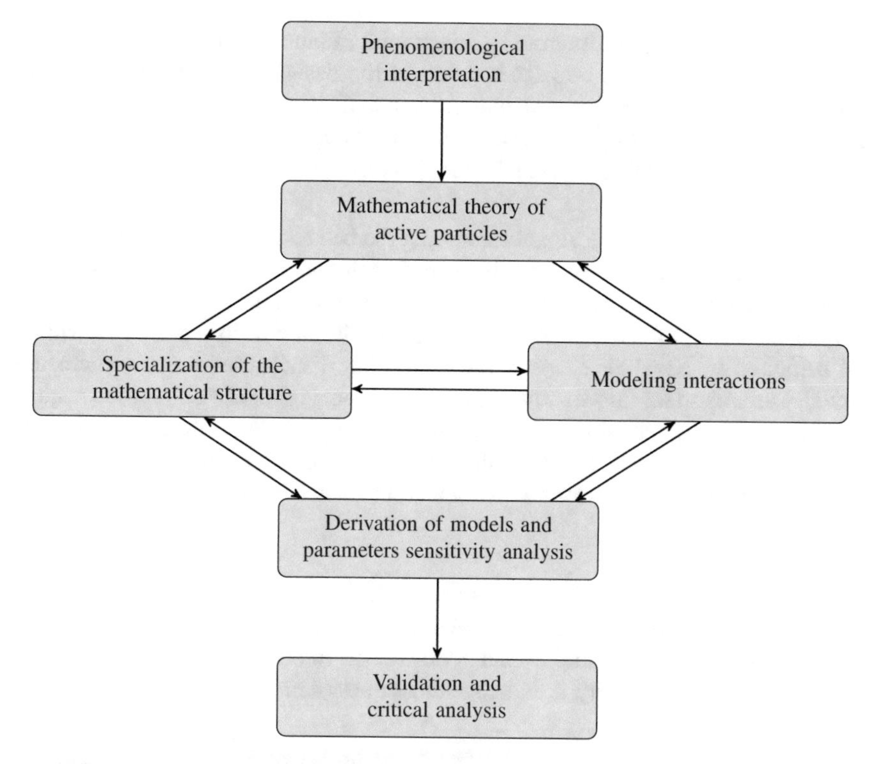

Fig. 4.2 Rationale toward the derivation of models

6. **Validation of models:** *A detailed analysis of the predictive ability of the model to depict quantitatively empirical data, when available, and qualitatively observed emerging behaviors.*
7. **Critical analysis toward improvements:** *The output of all preceding steps is object of a critical analysis focused on possible improvements of the models. This can imply also possible refinements of the mathematical structure.*

A minimal model includes a number of parameters equal to the number of different types of interactions. Reducing the complexity of the model means reducing the number of interactions. Generally, this reduction amounts to reduce the predictive ability.

4.4 Additional Rationale on the Validation of Models

As already mentioned, models are derived by inserting the description of interactions at the microscopic scale into the mathematical structure deemed to support the derivation of models. Therefore, two main ingredients lead to their derivation, namely the mathematical structure and the model of interactions at the microscopic scale. This feature has to be kept in mind in the design of a validation strategy. The validation process should take into account that different types of models can be derived according to different types of predictive features they can offer. In particular, two main typologies of models can be identified:

Predictive models which aim at foreseeing the behavior in time of systems given appropriate initial conditions and, if needed, constraints on the solutions. These models should reproduce, qualitatively and quantitatively, the dynamics by solution of mathematical problems. Therefore, prediction limited to low-order moments is not sufficient as the whole shape of the probability distribution over the microstate provides an important information. In addition, models should also reproduce qualitatively emerging behaviors which might be subject to large deviations, but are often qualitatively repeated within suitable variations of the parameters of the mathematical problem. Comparisons with empirical data need a proper assessment of the parameters of the model.

Explorative models which aim at exploring the behavior in time of systems under programmed variation of the parameters and of external actions. In fact, in the case of open systems, the role of external actions need to be studied to verify whether these actions have the ability, or not, to address the system toward a desired behavior.

In general, validation of models refers to the ability of models to reproduce the dynamics of emerging behaviors and their dependence on the variation of parameters and on the action of the external environment. In addition, validation can refer to a general predictive ability models that, after validation, can be used also as explorative ones. Once a model is derived, simulations might put in evidence emerging behaviors that are not yet observed. This event should suggest further search of empirical data that might, or not, confirm the presence of the aforementioned behavior.

An important feature of the validation process consists in verifying the ability of models to capture, as far as it is possible, the complexity features of the system object of the modeling. Subsequently, one can explore their ability to reproduce "quantitatively" available empirical data and "qualitatively" collective emerging behaviors which are observed by experiments.

A very special case of emerging behavior is the appearance of non-predictable events with the characteristics of the so-called *black swan*. A model might hopefully put in evidence this event for certain choice of the parameters of the model. Further, the model might put in evidence tipping points [23] which anticipate the arrival of this particular onset.

It is not an easy task as the amount of empirical data to develop a detailed validation process is quite limited in the case of complex systems. Moreover, these data, when available, correspond to steady uniform states and, in some cases, equilibrium, while models should operate far from such states.

An additional difficulty is that the greatest part of empirical data are available at the macroscopic scale, while the modeling process needs a detailed understanding of the dynamics at the microscopic scale. Therefore, a validation strategy should be elaborated to exploit the existing data offered by empirical data at the best of the panorama they offer.

Bearing all above in mind, let us define more precisely a validation strategy according to the following requirements that a "valid" model should fulfill:

1. Ability to capture the complexity features of the system object of the modeling approach.
2. Reproduction, at a quantitative level, of empirical data whenever available. These data generally correspond to steady conditions. Therefore, this condition is only necessary, but not yet sufficient.
3. Empirical data should not be artificially inserted into the model as, unfortunately, happens in various known models.
4. Models should qualitatively reproduce the emerging collective behaviors which, in some cases, can subject to large deviations, but in several cases preserve the qualitative behavior.
5. A detailed analysis should be developed to understand how these deviations can break up the qualitative shapes of the said patterns.
6. Models should take into account all physical conditions that can determine different observable dynamics. Each physical feature should be linked to a parameter directly related to them.
7. Derivation of models at the macroscopic scale from the underlying description delivered by the kinetic theory approach should be compared with the models obtained by the direct macroscopic approach.

Some additional remarks aim at a further enlightening of the complex strategy, presented in the above items.

- **Item (1)** corresponds to the validation of the mathematical structure needed for the derivation of models. Namely, the structure should show the ability to capture the complexity features that, even if heuristically, appear relevant for a certain class of system.
- **Items (2 and 3)** can be operative if and only if the first step has been fully investigated. Then, if empirical data are available, the model should reproduce them. However, since in most cases these data correspond to steady situation, this step must be viewed, as mentioned in Item 2, a necessary but not sufficient condition for the validation.
- **Items (4 and 5)** complete the validation by showing its ability to reproduce qualitatively observed emerging behaviors. Particularly interesting is the analysis of large deviations that might correspond to breaking of the observed patterns.
- **Item (6)** is rather a technical requirement that refers to the applicability of the model to real physical situations, where all features which appear in the system and which can modify the dynamics have to be inserted in the model.

- **Item (7)** corresponds to a qualitative information which refers a model obtained from the kinetic theory approach to the corresponding model derived by the approach of continuum mechanics. Some discrepancies can be found, which induce to further investigation on the outputs of both approaches and, if the case, promote suitable revisions.

An additional difficulty that appears in all systems is that the external environment can have an important influence on the dynamics. Therefore, appropriate models, and parameters, are needed to account for this type of actions. Interaction with the external environment can generate mutations and selections which might end up with the strengthening or weakening of functional subsystems.

We are aware that this general rationale might appear obscure to the reader as they are not related to specific case studies. Therefore, we will bring, in the next chapters, to the reader's attention various examples that show how the general ideas of this section can be finalized to well-defined applications.

4.5 Critical Analysis

This chapter has presented a conceptual bridge between the theoretical tools of Chapter 3 and the applications proposed in the next chapters. Two main topics have been treated, namely the search for a modeling strategy, where it is shown how the general structure derived in Chapter 3 can lead to the actual derivation of models and their validation by an appropriate strategy. Their operative application will be made plain in the next chapters, where the key issue is the modeling of interactions at the microscopic scale.

Some general ideas have been proposed to validate models. Intuitively, one might figure out that when a specific mathematical structure has been specialized from the general one to each well-defined class of systems, such a structure is valid if it has the ability to capture the most important complexity features of the system object of the modeling approach.

Moreover, it has been mentioned above that if interactions at the microscopic scale are properly inserted into such a structure, the model is expected to reproduce quantitatively empirical data and qualitatively emerging behaviors. In addition, models should be able to describe different scenarios of the dynamics depending on the variable conditions of the environment where the dynamics occurs.

It is worth stressing that modeling and validation are strictly related. In fact, validation can lead, as we have seen, to further improvements in both models and mathematical structures.

A final remark is that a model and the generating structure should not viewed as the final step of a process, but an intermediate step is to be critically analyzed in view of continuous process of revisions and improvements.

The applications that will be treated in the next chapter will be developed also with the aim of enlightening how all conceptual difficulties critically analyzed in this chapter can be properly treated.

Finally, let us stress that the selection of the two case studies pursues the objective of making the reader aware of all details of the application of the mathematical approach. Then, Chapter 7 proposes some speculations toward the development of a mathematical theory of soft sciences.

Chapter 5
Modeling Social Behavioral Dynamics

5.1 Plan of the Chapter

This chapter shows how the mathematical tools presented in Chapter 3 can be developed to derive mathematical models in social sciences and to understand how interactions at the microscopic scale can lead to the collective social behaviors.

The interest in behavioral social and economical sciences is rapidly growing with the aim of explaining the onset and consequences of sudden changes, including situations of economical crisis, such as those which appeared in our society in recent years. This growing interest motivates the contents of this chapter, where the following rationale guides the presentation:

- Interpretation of the phenomenology of the class of systems object of the modeling approach;
- Revisiting the general structure proposed in Chapter 3 by adjusting it to the specific contents of this chapter;
- An overview and critical analysis of what the literature offers on the modeling of social dynamics, focusing both on theory and applications;
- A presentation of conceivable developments towards a modeling approach to be organized within appropriate research programs;
- An overview on applications in fields different from that dealt with in this chapter.

The style of presentation will be followed also in the next chapter devoted to models that include a space structure. This style is consistent with the aim of this book where not only tools for applications are proposed, but also perspective research ideas are brought to the reader's attention.

The presentation strategy aims at showing how the approach proposed in our book can be applied in a well-defined case study, while a variety of applications are brought to the attention of the reader looking ahead to possible developments. The plan of the chapter is as follows:

N. Bellomo et al., *A Quest Towards a Mathematical Theory of Living Systems*, Modeling and Simulation in Science, Engineering and Technology, DOI 10.1007/978-3-319-57436-3_5

- Section 5.2 reviews the literature on social systems and motivates the modeling approach proposed in this chapter which focuses on various aspects of behavioral sciences and economy.
- Section 5.3 defines the rationale which guides the application of the tools developed in Chapter 3 to the modeling and analysis of social systems. A mathematical structure is derived to provide the conceptual basis for the modeling approach. This section also critically analyzes the existing literature and some specific contributions such as the modeling of wealth dynamics related to different welfare policies and progress or recession of nations [107].
- Section 5.4 proposes a new systems approach to social sciences, where mathematical tools are delivered by the methods presented in this book. The theoretical approach focuses on a specific case study concerning models of the onset and development of criminality in urban areas [35]. This dynamics can be related to the welfare policy.
- Section 5.5 proposes a variety of research perspectives focusing on analytic problems posed by applications of models to the study of real systems. In addition, this section presents a variety of possible modeling approaches to systems technically different from those treated in Section 5.4, whose modeling can, however, be developed according to guidelines analogous to those we have seen in the preceding sections.

A useful reference is paper [4], which can be considered a very first step toward the search of mathematical foundations of a systems approach, while the survey [5], which reports on more recent literature and developments, is a constant reference for this present chapter.

5.2 On the Contribution of Mathematics to Social Sciences

This section presents some reasonings on the interactions between mathematics and social sciences. First we look for the conceptual motivations related to systems, where individual behaviors can have an important impact on the overall social and economical dynamics. Subsequently, some alternative approaches, extracted from the literature on complex systems, are critically analyzed. These topics are treated in the next two subsections.

5.2.1 From bounded rationality to behavioral social dynamics

The motivation to apply the mathematical tools proposed in Chapter 3 to model complex social systems can be found in the radical philosophical change developed in the study of social and economic disciplines in recent years. Quoting from [6]:

> **A reconciliation among Economics, Sociology, and Psychology has been taking place, thanks to new cognitive approaches toward Economics in general [15, 108, 161]. New**

branches of Economics, much more linked to Sociology and Psychology than it used to be, are emerging.

Starting from the concept of bounded rationalities [222, 223], critiques to the traditional assumption of rational collective behavior [128, 149, 231] lead to the idea of Economics as a discipline highly affected by either rational or irrational individual trends, reactions, and interactions.

This innovative point of view promoted the idea of Economy as an evolving complex system [15], where heterogeneous individuals [46] interact to produce emerging unpredictable outcomes. In this context, the development of new mathematical descriptions, able to capture the complex evolving features of socioeconomic systems, is a challenging, though difficult, perspective, which calls for a proper interplay between mathematics and social sciences.

An important goal consists in the assessment of theoretical paradigms, and related mathematical structures, which can act as a conceptual background for a unified mathematical approach to the modeling of social systems characterized by common complex features.

The analysis of the complexity features of living systems has been already treated in the first two chapters. This chapter specializes it to the case of social systems. The following features can be briefly mentioned thus revisiting the general concepts which were previously introduced.

- Living entities have the ability to develop specific strategies to fulfill their well-being depending on their own state and on that of the entities in their surrounding environment.
- Individual behaviors are heterogeneously distributed and can play an important role in determining the overall collective dynamics, especially when irrational behaviors of a few entities can induce large deviations from the usual dynamics observed in rationality-driven situations.
- Interactions are nonlinearly additive and involve immediate neighbors, but in some cases also distant individuals. Interactions promote the ability of learning from past experience. As a result, a continuous adaptation to the changing-in-time environmental conditions occurs.
- Interactions occur also across networks and involve both individuals and groups sharing the same interest. These interactions can have an important influence on the overall dynamics of the system by changing the trends generated by interactions within the same node.
- Social systems can even undergo processes that include selection and evolution. For instance, aggregation of individuals can be viewed as the formation of groups of interest [4], namely aggregation of individuals which pursue a common objective. These groups can generate new groups better adapted to an evolving social and economical environment. These groups might increase their presence, while other groups might disappear for the opposite reason.
- In some cases, large deviations break out the macroscopic (qualitative) characteristics of the dynamics, whence substantial modifications can be observed. These deviations can be interpreted as the requisites for events that can be classified as

a "Black Swan" [230]. Particularly important is the detection of early signals to predict the onset of such sudden large deviations [23].

5.2.2 *Mathematical tools toward modeling social dynamics*

Social systems are prominent examples of complex systems and, in recent years, a growing interest has arisen to study their collective emerging behaviors such as cooperation, cultural conflicts, and problems of social consensus.

Two complementary ways of modeling social systems can be roughly distinguished.

The first approach involves the design and analysis of simplified mathematical models that do not try to mimic the real behavior of individuals but abstract the most important qualitative aspects so as to gain insight into the system dynamics. The rationale behind this approach is that, in analogy with the "universality" concept in statistical physics, certain aspects of complex behavior are supposed to be independent on the specific dynamical details.

The second approach consists in designing more comprehensive and realistic models, usually in the form of numerical simulations, which represent the interacting parts of a complex system, often down to minute details. The border between these two approaches is not sharply defined, and tools of each of them are always more often applied together.

The first approach has been mainly adopted to investigate social systems using methods of the statistical physics. Two research fields have been rapidly grown referring specifically to as econophysics [247] and sociophysics [78, 121, 241]. Although the principles of both fields have much in common, econophysics focuses on the narrower subject of economic behavior while sociophysics studies a broader range of social issues, including social networks, language evolution, population dynamics, epidemic spreading, terrorism, voting, coalition formation, and opinion dynamics.

A more realistic description of real-world social interactions can be build within game-theoretical models. In its original formulation, game theory has been used to analyze situations of conflict or cooperation in economics [176, 183]. Two major streams of extensions of classical game theory have been developed as early as in the sixties of the last century, namely, differential games [212] and games with incomplete information [136]. The former are characterized by continuously time-varying strategies and payoffs with a dynamical system governed by ordinary differential equations, while in the latter players are supposed to ignore information about the others players for what strategies and payoffs are concerned.

An important breakthrough of game theory is behavioral game theory, also called experimental game theory [76, 129]. The main idea is that, instead of analyzing games theoretically, experimenters get real people to play them and record results. The remarkable finding is that in many situations people respond instinctively and play according to heuristic rules and social norms rather than adopting the strategies indicated by rational game theory.

Evolutionary game theory is an extension of the classical paradigm toward this concept of bounded rationality. Unlike classical game theory, the focus is on large populations of individuals who interact at random, rather than on a small number of individuals who play a repeated game and individuals are assumed to employ adaptive rules rather than being perfectly rational. It is worth noticing that behavioral rules are dynamic rules, specifying how much players take into consideration the game's earlier history and how long they would think ahead. A static, equilibrium-based analysis turns out to be not sufficient to provide enough insight into the long-run behavior of such systems [154].

Topics in economics studied using the methods of evolutionary game theory range from behavior in markets [235], to bargaining [200] to questions of public good provision and collective action [181]. Applications to problems of broader social science interest include residential segregation [250], cultural evolution [66], and the study of behavior in transportation [224]. In addition, it is worth noticing that evolutionary game theory overlaps with another active area of research in mathematical biology, namely population ecology and population genetics [240] as explicitly discussed in [154].

Although behavioral rules control the system at the microscopic, individual level, evolutionary game theory in its standard form considered population dynamics on the aggregate level. However, in practical applications, it may be important to properly account for individual preferences, payoffs, strategy options (heterogeneity in agent types), and/or specific local connection among individuals (structural heterogeneity) [229]. The systematic investigation of such heterogeneities requires a change of perspective in the description of the system from the aggregate level to the individual level. The resulting huge increase in the relevant system variables makes most standard analytical techniques, operating with differential equations, fixed points, etc., largely inapplicable and prompted the development of agent-based models.

Agent methods are computational models which simulate the actions and interactions of autonomous agents (both individual or collective entities such as organizations or groups) with the aim to assess the collective emerging behavior. Since the very simple agent-based model on racial segregation proposed by Schelling [219], a plethora of models have been developed which have taken into account several features such as bounded rationality, personal well-being, wealth and social status/education, and many others [16, 108].

A recent trend is emerging with the aim of providing a unified approach, suitable to link methods of statistical mechanics and game theory. A first approach is due to Helbing, who has the great merit of pointing out that individual interactions need to be modeled by methods of game theory [138, 140]. Another approach is provided by the mean field games, where interactions among individuals are evaluated by supposing that each individual contributes to the creation of a mean field which then acts back on each individual [167, 168]. Both approaches mainly focus on the continuum limit of infinitely many individuals. The strong similarity between the mathematics of statistical physics and games where players have limited rationality has been recently pointed out in [246].

Methods of classical kinetic theory have been also applied to model various social systems from the pioneer papers [60, 61]. Nowadays, the most important reference is given by the book by Pareschi and Toscani [197], which provides an exhaustive overview covering the whole path from numerical methods to mathematical tools and applications, such as opinion formation and wealth dynamics, as well as various aspects of the social and economical dynamics of our society.

Methods of kinetic theory of active particles are the central topic of this chapter and represent, to the authors' opinion, the most promising approach to the modeling of social systems. As it has been previously discussed, the key feature is that the overall state of the system is described by a probability distribution over the microscale state, while interactions are modeled by stochastic evolutionary games. In Subsection 5.4.3, it is shown how the dynamics of criminality can be modeled according to the kinetic theory of active particles.

5.3 On the Search of a Mathematical Structure

This section shows how the general mathematical structure derived in Chapter 3 can be specialized to capture the specific features of social systems. The contents are proposed through four subsections: Subsection 5.3.1 provides a general description of a large class of social systems object of the modeling approach and outlines the modeling strategy which generates the mathematical structure we look for. Then, the representation of the system by linking at each functional subsystem a probability distribution over the microscopic state is defined. Subsection 5.3.2 derives the aforementioned mathematical structure. Subsection 5.3.3 proposes some reasonings on the modeling of interactions. Finally, an overview of models and their conceptual basis are presented in Subsection 5.3.4.

As already mentioned, most of the concepts given in this section simply particularize the general description proposed in Chapter 3 adapting it to the specific class of systems treated in this present chapter.

5.3.1 Heuristic description and representation of social systems

Let us now define precisely the class of systems which is object of the modeling approach based on the mathematical tools proposed in this book:

- The entities which comprises the system are referred to as active particles and are assumed to be distributed in a network of nodes labeled by the subscript $i = 1, \ldots, n$;

- The role of space variable is confined to the network, namely a continuous variation in space is replaced by migration dynamics across the nodes;
- Active particles are partitioned into *functional subsystems*, where they collectively develop a common strategy. Functional subsystems, in each node, are labeled by the subscript $j = 1, \ldots, m$. Therefore, the total number of subsystems is $N = n \cdot m$;
- The strategy is heterogeneously distributed among active particles and corresponds to an individual state, defined *activity*. The state of each functional subsystem is defined by a probability distribution over the said activity variable;
- Active particles interact within the same functional subsystem as well as with particles of other subsystems and are acted upon the external actions;
- Interactions are generally nonlinearly additive and are modeled as *stochastic games*, meaning that the outcome of a single interaction event can be known only in probability;
- This system can be viewed as an *internal system*, namely the network where $n \cdot m$ functional subsystems are localized, and can be subjected to $n \cdot m$ external actions upon over the active particles of the internal system.

These features are quite general, hence, addressing them more precisely to the modeling of real social systems is necessary. In particular, the following issues need to be accounted for when specific models are derived:

i) Localization of the so-called functional subsystems and dynamics across nodes;
ii) Transitions across functional subsystems and their characterization;
iii) Presence of leaders, which should be grouped in specific functional subsystems;
iv) Characterization of the so-called external actions and of the dynamics induced by them;
v) Detailed analysis of all type of interactions.

The reader interested to a deeper understanding of the mathematical theory of networks is addressed to a broad literature recently developed on this topic, for example [1, 21, 24, 123, 124, 236]. This chapter simply considers networks as possible localization of functional subsystems.

The overall state of the internal system, according to such approach, is delivered by a probability distribution over the activity variable in each node and for each functional subsystem, hereinafter denoted by the acronym ij-FS, while the generic one is simply denoted by FS.

In more detail, the probability distribution is denoted by

$$f_{ij}(t, u) \; : \; [0, T] \times D_u \; \rightarrow \; \mathbb{R}_+, \tag{5.1}$$

so that moments can provide a description of quantities of the system at the macroscopic scale, under suitable integrability conditions, as reported in Section 3.2, see Eqs. (3.3)–(3.9).

The actions applied by the *external system* are supposed, as we have seen in Subsection 3.3.4, to act at the microscopic scale, by *agents*, at the same scale of the system and over the same domain of the activity variable. These actions are assumed to be known function of space and of the activity variable defined by the following structure:

$$\varphi_{ij}(t,u) = \varepsilon_{ij}(t)\,\psi_{ij}(u) : \quad [0,T] \times D_u \to \mathbb{R}_+, \tag{5.2}$$

where ε_{ij} models the intensity of the action in the time interval $[0, T]$, while ψ_{ij} models the intensity of the action over the activity variable.

The activity is represented by scalar variable whose domain of definition D_u is typically $[-1, 1]$ or even the whole real axis for probability distributions which decay to zero at infinity. Positive values indicate an expression coherent with the activity variable, while negative values expressions in opposition. The interval $[-1, 1]$ is used whenever a maximum admissible value of the activity variable can be identified so that u is normalized to such value.

More in general, the activity variable can be a vector, but we remain here within the relatively simpler case of a scalar variable so that the formalism of the mathematical approach is not too heavy.

Generally, D_u is supposed, for simplicity, constant. However, in a more general approach, it can be assumed to be depending on the distribution function. A possible example is that of [52] based on the conjecture that sensitivity is related to a critical size rather than to the whole perception domain.

5.3.2 *Mathematical structures*

The mathematical tools proposed in Chapter 3 lead to the derivation of structures, which describe the evolution in time of the probability distribution over the activity variable for each functional subsystems of a large system of multi-agents that can be located in a network. The dynamics refer to the variables, defined in the preceding paragraph, which are deemed to describe the state of social systems.

The result, which consists in a system of $m \times n$ differential–integral equations, is obtained by a balance of particles within the elementary volume $[u, u + du]$ of the space of the microscopic states.

This approach has been already reported in Chapter 3, as well as in the already cited survey [47]. Therefore, only a concise description of the technical aspects of the approach will be given here.

- The dynamics of interactions can be obtained by assigning to each particles of the internal ij-FS the following distinguished roles: *Test particle* of the ij-th FS with activity u, at time t; *Candidate particle* of the pq-th FS with activity u_*, at time t, which can gain the test state u as a consequence of the interactions; *Field particle* of the hk-th FS with activity u^*, at time t, which triggers the interactions of the candidate particles.

- The overall state of the system is delivered by the probability distribution linked to the test particle.
- Interactions at the same or different scales are considered. More specifically, *microscopic interactions* of particles within the same functional subsystem or across them and *microscopic–macroscopic interactions* which, instead, involve particles with functional subsystems viewed as a whole being represented by their mean value.
- The mathematical framework describing different kind of interactions will be specified, for each type of interactions, by means of two terms: The *interaction rate* which describes the rate of interactions involving particles; The *transition probability density* which describes the probability density that a candidate particle falls into the state of the test particle after undergoing one of the interactions, where transitions can occur within the same node and functional subsystems, as well as across them.
- The derivation of the mathematical framework is obtained, as in Chapter 3, by a balance of number of particles in the elementary volume of the state space.

In more detail, the balance flux mentioned in the last item corresponds to the following:

Dynamics of the number of active particles
= *Net flux due to ms-interactions between a-particles*
+ *Net flux due to mMs-interactions between a-particles and f-subsystems*
+ *Net flux due to ms-interactions between a-particles and external actions*,

where the net flux denotes the difference between inlet and outlet fluxes, while “ms” stands for microscopic scale, “mMs” stands for microscopic–macroscopic scale, “a-particles” for active particles, and “f-subsystems” for functional subsystems.

The result is the following structure:

$$\partial_t f_{ij}(t, u) = A_{ij}[\mathbf{f}](t, u) + B_{ij}[\mathbf{f}](t, u) + C_{ij}[\mathbf{f}, \varphi](t, u), \tag{5.3}$$

where $\mathbf{f}$ denotes the set of all distribution functions and

$$\begin{aligned} A_{ij}[\mathbf{f}](t, u) = & \sum_{p,h=1}^{n} \sum_{q,k=1}^{m} \int_{D_u \times D_u} \eta_{pq,hk}[\mathbf{f}](u_*, u^*)\, \mathscr{A}^{ij}_{pq,hk}[\mathbf{f}](u_* \to u | u_*, u^*) \\ & \times\, f_{pq}(t, u_*)\, f_{hk}(t, u^*)\, du_*\, du^* \\ & -\, f_{ij}(t, u) \sum_{h=1}^{n} \sum_{k=1}^{m} \int_{D_u} \eta_{ij,hk}[\mathbf{f}](u, u^*)\, f_{hk}(t, u^*)\, du^*, \end{aligned} \tag{5.4}$$

$$
\begin{aligned}
B_{ij}[\mathbf{f}](t,u) = & \sum_{p,h=1}^{n}\sum_{q,k=1}^{m}\int_{D_u} \nu_{pq,hk}[\mathbf{f}](u_*, \mathbb{E}_{hk}^1[f_{hk}]) \\
& \times \mathscr{B}_{pq,hk}^{ij}[\mathbf{f}](u_* \to u | u_*, \mathbb{E}_{hk}^1[f_{hk}])\, f_{pq}(t,u_*)\, du_* \\
& - f_{ij}(t,u) \sum_{h=1}^{n}\sum_{k=1}^{m} \nu_{ij,hk}[\mathbf{f}](u, \mathbb{E}_{hk}^1[f_{hk}]),
\end{aligned} \tag{5.5}
$$

and

$$
\begin{aligned}
C_{ij}[\mathbf{f}](t,u) = & \sum_{p=1}^{n}\sum_{q=1}^{m}\int_{D_u\times D_u} \mu_{pq}[\mathbf{f},\varphi](u_*, u^*)\, \mathscr{C}_{pq}^{ij}[\mathbf{f},\varphi](u_* \to u | u_*, u^*) \\
& \times f_{pq}(t,u_*)\, \varphi_{pq}(t,u^*)\, du_*\, du^* \\
& - f_{ij}(t,u) \int_{D_u} \mu_{ij}[\mathbf{f},\varphi](u,u^*)\, \varphi_{ij}(t,u^*)\, du^*.
\end{aligned} \tag{5.6}
$$

Interaction rates and the transition probability densities have been denoted by $\eta_{pq,hk}$, $\nu_{pq,hk}$, μ_{pq} and $\mathscr{A}_{pq,hk}^{ij}$, $\mathscr{B}_{pq,hk}^{ij}$, $\mathscr{C}_{pq}^{ij}$, respectively, while, as mentioned, external actions $\varphi_{ij}(t,u)$ are assumed to be known functions of time and of the activity variable. Square brackets have been used in Eqs. (5.3)-(5.6) to denote the dependence on the distribution functions which highlight the nonlinear nature of interactions.

We refer to Chapter 3, Sections 3.3 and 3.4, to enlighten some specific features of this structure and possible generalizations. In particular, the remarks in what follows have been already presented in the more general context of living systems, however, they are here addressed, with some obvious repetitions, to the class of models treated in this chapter.

1. It has been assumed that each node includes all FSs. This assumption is formal if the initial status of some FSs is equal to zero and remains like that for $t > 0$. On the other hand, such status can take values for $t > 0$ due to migration phenomena.
2. This structure includes some specific systems. As an example, the case of a closed system can be considered by setting $\varphi = 0$. The structure simplifies as follows:

$$
\partial_t f_{ij}(t,u) = A_{ij}[\mathbf{f}](t,u) + B_{ij}[\mathbf{f}](t,u). \tag{5.7}
$$

3. A possible alternative to the structure can be obtained by modeling an external action $K_{ij}(t,u)$ acting over each functional subsystem, which yields:

$$
\partial_t f_{ij}(t,u) + \partial_u\big[K_{ij}(t,u) f_{ij}(t,u)\big] = A_{ij}[\mathbf{f}](t,u) + B_{ij}[\mathbf{f}](t,u), \tag{5.8}
$$

or by supposing that the external actions act over the system by the mean value of ψ_{ij} so that the term C_{ij} is modified accordingly. Further generalizations can be achieved by taking into account also actions at the macroscopic scale by each node as a whole and by the whole network.

4. Several applications, for example [6, 46], suggest to use discrete variables at the microscopic scale. In fact, in some specific cases, the state of the active particles is more precisely identified by means of ranges of values rather than by a continuous variable. Technically, this means that the distribution functions, for each functional subsystem, have to be regarded as discrete distribution functions, while the interaction terms map discrete variables rather than continuous ones. The mathematical structures are readily obtained by specializing the integrals over the microscopic states as sums over discrete states. Detailed calculations are reported in Subsection 3.4.1 and referring to social system in [5].
5. The mathematical structure (5.3)-(5.6) is based on the assumption that interactions occur in the whole domain D_u of the activity variable, and some applications require that interactions are restricted in a subdomain $\Omega \subseteq D_u$. If Ω is known, the integration domain has to be modified accordingly. See Subsection 3.3.5 for explicit calculations of the effective interaction domain.

5.3.3 Some reasonings toward modeling interactions

Specific models can be obtained by inserting, into Eq. (5.3), the description of interactions. Various recent papers have contributed to this topic. More in detail [35, 46, 107] have shown how interactions can be modeled by games where the output of the interactions is conditioned not only by the state of the interacting entities, but also by the probability distribution over these states. A theoretical analysis of the different sources of nonlinearity and their related mathematical structures is proposed in [36]. Models of multiscale micro–macro interactions have been proposed in [162] concerning the modeling of migration phenomena, in [163] on opinion formation in small networks, and in [35] focused on criminality dynamics. These multiscale features deserve attention in several fields of social dynamics.

Although the mathematical description of interactions can be precisely developed only if addressed to specific models, some general rationale can be extracted. Let us consider, separately according to such aim, the modeling of the interaction rates and of the transition probability density.

Interaction rate: The modeling of the interaction rate requires the definition of different concepts of *distance*. The following ones can be defined:

(i) *Microscopic state distance* $|u_* - u^*|$ refers to the states at the microscopic scale between candidate and field particles;
(ii) *Microscopic–macroscopic distance* $|u_* - \mathbb{E}[f_{pq}]|$ refers to the interaction of a candidate particle with state u_* with the group of particles belonging to a FS with mean value of the activity $\mathbb{E}[f_{pq}]$;
(iii) *Affinity distance* refers to the interaction between active particles characterized by different distribution functions, where this distance is introduced according to the general idea that two systems with close distributions are *affine* and is given by

$||f_{pq} - f_{hk}||$, where $|| \cdot ||$ is a suitable norm to be chosen depending on the physics of the system.

In general, the interaction rate decays with the distance, where heuristic assumptions lead to a decay described by exponential terms or rational fractions. An immediate example is the following:

$$\eta_{pq,hk}[\mathbf{f}](u_*, u^*) = \eta_0 \exp\{-|u_* - u^*| - c_1 \, |u_* - \mathbb{E}[f_{pq}](t)| - c_2 \, ||f_{pq} - f_{hk}||\}, \tag{5.9}$$

where η_0, c_1, and c_2 are positive defined constants. This equation is simply a technical modification of Eqs. (3.24)-(3.25).

Transition probability density: The dynamics of interactions can be modeled by theoretical tools from evolutionary and behavioral game theory [76, 130, 140, 186, 214, 215, 218], which provides features to be introduced into the general mathematical structure in order to obtain specific models. Additional tools are given by learning theory [75] and evolutionary games [186], within the broader context of statistical dynamics and probability theory. Various types of games have already been defined in Subsection 3.3.1. Here we simply add a few remarks useful to enlighten their use in the modeling of social systems.

In *competition games*, the advantage is for only one of the two players involved in the game, while in *cooperation games* the interacting particles show a trend to share their microscopic states due to a sort of attraction effect [146, 177]. *Learning interactions* introduce a partial cooperation as one of the two particles modifies, independently of the other, its microscopic state, while the other attempts to reduce the distance by a learning process [75]. *Hiding/chasing interactions* introduce again a sort of competition as one of the two particles attempts to increase the distance from the state of the other one (*hiding*), which conversely tries to reduce it (*chasing*) [35]. All these dynamics can be repeated in *micro–macro interactions* which occur between a particle with the mean value of each functional subsystem, but the mean state of the FSs is not modified by the interactions, but it follows the dynamics of the whole system [25, 204].

Let us also remind that all aforesaid types of games can occur, as mentioned in Chapter 3, simultaneously in a general context of heterogeneous particles. In some cases, see [46], the occurrence of one of them is ruled by a *threshold* on the distance between the states of the interacting particles. Such a threshold is, in the simplest case, a constant value.

A recent paper [107] has shown that the threshold can depend on the state of the system as a whole and that can have an important influence on the overall dynamics. More in detail, an excess selfishness in the wealth distribution ends up with an overall reduction of the total wealth, which is shared only by small groups, while the great part of the population is confined in poverty [107]. A deep insight into the role of selfishness and the complex problem of its "measurement" is proposed in [221].

This complex dynamics generally refers to the interaction of more than one type of dynamics as treated in [63] from the view point of social economy, where political choices are related to economical growth. Further studies on this topic, in addition to the already cited paper [107], are proposed in [46] and [35].

5.3.4 An overview of models and their conceptual basis

The mathematical structure proposed in Subsection 5.3.2 has been already used, in a variety of models known in the literature. Therefore, it is worth developing an overview of these particular models to understand which type of innovation can be achieved by a more general approach. This analysis can contribute to enlighten how the various approaches are related to the complexity features of social systems.

• *Models with linear interactions:* The application of kinetic-type equations with interactions modeled by theoretical tools of game theory has been introduced in [60] for large social systems with discrete states at the microscopic scale. The authors used games with linear binary interactions and derived a system of ordinary differential equations suitable to describe the time dynamics of a discrete probability distribution corresponding to the discrete microstates. Actually, the authors used a simplified version of a framework proposed in [41] to model complex multicellular systems. The application of [60] was referred to the dynamics of wealth distribution and was based on consensus/dissent games triggered by a constant threshold of the distance between the states of the interaction pairs.

The same approach was subsequently extended, with minor technical modifications, to various types of applications by the authors and their coworkers. Without claim of completeness, examples include opinion formation [61], taxation dynamics with some social implications [62], and resilience [95]. Various innovative developments followed these pioneer papers. The main sequential steps of the aforementioned developments are summarized in the following.

• *On the introduction of nonlinear microscopic interactions:* Nonlinearity for interactions at the microscale has been introduced in [46] and [107]. Both papers analyze the interplay of different dynamics, namely wealth distribution versus support or opposition to Governments, or selfishness in wealth dynamics as a source of overall wealth dissipation in nations. The role of nonlinearity of interactions was clearly put in evidence. Interesting reference for social aspects of selfishness and wealth distribution is given in [166, 221].

• *Dynamics over small networks:* The modeling of this type of dynamics by kinetic theory methods was introduced in [162] and [163] referred to specific applications, as an example migration phenomena. These two papers have introduced a new type of nonlinearity under the assumption that individuals are sensitive to the mean value, which depends on the distribution function, of each functional subsystem. Indeed, this is the two-scale micro–macro interaction presented in our book. The concept of social distance is one of the main tools toward the study of networks. Such a distance can be viewed as a stochastic quantity as it depends on a probability distribution. Therefore, the social network evolves in time and is different from the physical one.

• *Mutations and selection:* A post Darwinian dynamics consisting in mutations followed by selection plays an important role in biology, specifically in the immune competition [57, 91]. An analogous dynamics appears in socioeconomic systems, where new groups of interest can be generated in the dynamics, for instance by aggregation of different groups, and subsequently they might either expand or disappear in a competition somewhat mediated by the environment where these groups act. These concepts have been introduced in some recent papers, for instance [89], in a behavioral theory of urbanism. However a systematic approach to social sciences in a mathematical framework still needs to be exhaustively developed.

5.4 Hallmarks of a Systems Theory of Social Systems

This section shows how the mathematical tools presented in Section 5.3 can be applied to modeling and simulation of social systems. In particular, we aim at showing how a general systems theory can be developed. The contents are presented through four subsections focused on the following topics: ability to capture the complexity features of social systems; hallmarks of the systems sociology approach; technical implementation of the theory to a specific case study, namely modeling the onset and propagation of criminality; simulations to analyze quantitatively the predictive ability of the model.

5.4.1 *Ability to capture complexity features*

A critical analysis on the ability of the mathematical structures presented in Section 5.3 which are deemed to capture the complexity features presented in Section 5.2 can be useful toward the design of the aforementioned systems approach. In details:

- *Ability to express a strategy* : The microscopic state includes the activity variable, which is deemed to model the strategy expressed by each functional subsystem.
- *Heterogeneity* : The heterogeneous behavior is accounted for by the description of the system by means of a probability distribution linked to each functional subsystem.
- *Nonlinear interactions* : Linear additivity has not been applied. Furthermore, we observe that an additional source of nonlinearity is related to non-locality of interactions of the communications between functional subsystems localized in different nodes of the network.
- *Learning and adaptation* : The ability of living systems to learn from past experience can be accounted for by modeling interactions based on rules that evolve in time and include a continuous adaptation to the changing-in-time environmental conditions.

- *Selection and evolution* : The onset of functional subsystems is modeled by the transition probability density, where interactions include the formation of groups of interest, which in turn can generate new groups more suited to an evolving social and economical environment. Selection is modeled by interactions between the internal and external system.

It is worth noticing that the descriptive ability is only potential, and it should be practically implemented in the derivation of models when theoretical tools of game theory are used to describe interactions.

Although models rapidly reviewed in Subsection 5.3.4 have been successful to depict some collective behaviors observed in reality, it can be remarked that one of the limits of the existing literature is that it deals with closed systems, while interactions with the outer environment have not been taken into account.

5.4.2 Towards a systems approach

The rationale toward a systems approach were proposed in [5]. These can be summarized as follows:

1. The domain of the space variable, where the system is located, is subdivided into interconnected *sub-domains* called *nodes*, while the set of all nodes and their connection is called *network*.
2. The individual entities, called *active particles*, are aggregated into different groups of interest called *functional subsystems*.
3. The active particles aggregated in a functional subsystem express a *common strategy* (or expression of interest) called *activity*. The set of all functional subsystems defines the *internal system*.
4. The set of all actions applied to functional subsystems defines the *external system*.
5. The activity of particles within each functional subsystem is heterogeneously distributed over the microscopic state, and it is defined by a *probability distribution*. The overall set of probability distributions represents the *dependent variable* of models.
6. Modeling interactions between active particles within the same functional subsystem and with particles of other subsystems, as well as *external actions*. Interactions can induce transitions across functional subsystems.
7. Derivation of the equations which define the dynamics of the probability distributions by a *balance of particles within elementary volumes of the space of microscopic states*, the inflow and outflow of particles being related to the aforementioned interactions.
8. *Solution of these equations by suitable computational methods* to obtain both the probability distributions and the moments suitable to provide an information on macroscopic states.

Let us finally remark that following specific modeling actions and variables: *Partition, activity variables, actions, network, and so on* are not constant features of the same overall system. In fact, these variables depend on the specific social dynamics object of study. Therefore, different investigations would lead different characterization of the overall system and hence of mathematical models. Further, the role of networks in determining the overall dynamics has been treated only tangentially. Therefore, a great deal of work is still waiting for being developed. Hopefully, the next sections will contribute to link some of the conceptual ideas proposed until now to a real application.

5.4.3 Reasonings on a case study

Let us now show how the general systems approach outlined in the preceding section can be applied in a specific case study. In detail, we consider the onset and development of criminality in a society, where this dynamics is contrasted by intelligence and police forces.

The contents are referred to a mathematical literature, which is constantly growing due also to the impact that this topic has on the well-being and security of citizens. Useful references are given by the following essays [59, 109, 110], which are brought to the attention of the reader as possible examples.

This subsection illustrates how the hallmarks of the system theory can be technically applied to the model presented in [35], which shows how the contrast actions are related to one side on the social state of citizens, such as wealth and culture, and, to the other side, to the ability of security forces based also on their training.

More in general, the interplay between wealth distribution and unethical behaviors is object of growing interest of researchers, who operate in the field of social sciences [166], while it is becoming a new field of investigation of applied mathematicians [46, 107].

The selected case study is limited to closed systems. However, we will discuss how the approach can be generalized to open systems. The model accounts for the following interaction dynamics:

- Susceptibility of citizens to become criminals;
- Susceptibility of criminals to reach back the state of normal citizens;
- Learning dynamics among criminals;
- Motivation/efficacy of security forces to catch criminals;
- Learning dynamics among security forces, and interplay with wealth dynamics.

In more details, the following features of the model proposed in [35] are briefly analyzed:

Network: Space dynamics is not accounted for and therefore the system is supposed to comprise only one node.

Functional subsystems and activity: The following subsystems are considered: citizens, criminals, and security forces, while the activity variables expressed by them are, respectively, the following: wealth, criminal ability, and detective ability.

Individual based interactions and encounter rates: Citizens with closer social states interact more frequently; experienced lawbreakers are more expected to expose themselves; experienced detectives are more likely to chase less experienced criminals.

Modeling interactions, which induce modification of state: Citizens are susceptible to become criminals, motivated by their wealth state. As an example, a candidate citizen interacting with a richer one can become a criminal, mutating into a new functional subsystem. Criminals interact among themselves resulting in a dynamics by which less experienced criminals mimic the more experienced ones. Moreover, also interactions with less experienced lawbreakers increase the level of criminality. Criminals chased by detectives are constrained to step back decreasing their activity value as the price to be paid for being caught. At the same time, detectives gain experience from catching criminals increasing their activity. Due to this action, criminals are induced to return to the state of normal citizens with probability which increases with decreasing values of their level of criminality and increasing values of skill of detectives.

Interactions between individuals and the whole functional subsystem: Both criminals and detectives interact with the mean value through the mean-micro state distance within their own functional subsystem; in both cases, the interaction rate increases with the distance between the individual state and the mean value. Criminals and detectives are subject to interactions also with their respective mean activity values. The dynamics is such that only those who are less experienced than the mean tend to learn and move toward it.

Let us now consider the mathematical formalization of the aforementioned concepts in a mathematical model.

Functional subsystems: The following functional subsystems and microstates are considered:

- $j = 1$, citizens with $u \in D_1$, wealth;
- $j = 2$, criminals with $u \in D_2$, criminal ability;
- $j = 3$, detectives with $u \in D_3$, experience/prestige.

The equations in the following suppose $D_1 = D_2 = D_3 = D_u = [0, 1]$.

Mathematical structure: The mathematical structure is obtained by a specialization of Eqs. (5.3)-(5.6). In detail:

$$\partial_t f_j(t, u) = A_j[\mathbf{f}](t, u) + B_j[\mathbf{f}](t, u), \tag{5.10}$$

where

$$A_j[\mathbf{f}](t,u) = \sum_{q,k=1}^{3} \int_{D_u \times D_u} \eta_{qk}(u_*, u^*) \mathscr{A}_{qk}^{j}(u_* \to u | u_*, u^*) f_q(t, u_*) f_k(t, u^*) \, du_* \, du^*$$
$$- f_j(t,u) \sum_{k=1}^{3} \int_{D_u} \eta_{jk}(u, u^*) \, f_k(t, u^*) \, du^*, \tag{5.11}$$

and

$$B_j[\mathbf{f}](t,u) = \int_{D_u} \nu_{jj}(u_*, \mathbb{E}_j^1) \mathscr{B}_{j,j}^{j}(u_* \to u | u_*, \mathbb{E}_j^1) f_j(t, u_*) du_*$$
$$- \nu_{jj}(u, \mathbb{E}_j^1) f_j(t,u). \tag{5.12}$$

The social structure of citizens is assumed to be fixed, namely, the time interval is supposed sufficiently short that the wealth distribution does not change in time. Interaction rates and transition probability densities are briefly summarized in the following (the reader is referred to [35] for more details).

Interaction rates: Different interaction rates characterize the different types of interactions:

For citizens interactions

- Interaction rate is higher for citizens with closer social states:

$$\eta_{11}(u_*, u^*) = \eta^0 \left(1 - |u_* - u^*|\right). \tag{5.13}$$

For criminals interactions

- Interaction rate between criminals is higher for experienced criminals:

$$\eta_{22}(u_*, u^*) = \eta^0 (u_* + u^*). \tag{5.14}$$

- Interaction rate between criminals and detectives is higher for unexperienced criminals and experienced detectives:

$$\eta_{23}(u_*, u^*) = \eta^0 \left((1 - u_*) + u^*\right). \tag{5.15}$$

- Interaction rate between criminals and the lawbreakers environment is higher for criminals whose ability is lower than the mean ability of criminals:

$$\nu_{22}(u_*, \mathbb{E}_2^1) = \begin{cases} 2\nu^0 |u_* - \mathbb{E}_2^1| & u_* < \mathbb{E}_2^1, \\ 0 & u_* > \mathbb{E}_2^1. \end{cases} \tag{5.16}$$

For detectives interactions

- Interaction rate between detectives and criminals is higher for experienced detectives and unexperienced criminals:

$$\eta_{32}(u_*, u^*) = \eta^0 \left(u_* + (1 - u^*)\right). \tag{5.17}$$

- Interaction rate between detectives and the security forces environment is higher for detectives whose ability is lower than the mean ability of detectives:

$$\nu_{33}(u_*, \mathbb{E}_3^1) = \begin{cases} 2\nu^0 |u_* - \mathbb{E}_3^1| & u_* < \mathbb{E}_3^1, \\ 0 & u_* > \mathbb{E}_3^1. \end{cases} \tag{5.18}$$

Transition probability densities: The same type of interactions are now considered with the aim of modeling the transition probability densities.

For citizens interactions

- Citizens are susceptible to become criminals, motivated by their wealth state. More in detail, a candidate citizen with activity u_* interacting with a richer one with activity $u^* > u_*$ can become a criminal, mutating into functional subsystem $j = 2$ with a very low criminal ability $u = \varepsilon \approx 0$. In particular, it is assumed that the transition probability increases with decreasing wealth:

$$\mathscr{A}_{1,1}^1(u_* \to u | u_*, u^*) = (1 - \alpha_1 (1 - u_*) u^*)\, \delta(u - u_*), \tag{5.19}$$

and

$$\mathscr{A}_{1,1}^2(u_* \to u | u_*, u^*) = \frac{1}{\varepsilon} \alpha_1 (1 - u_*) u^* \chi_{[0,\varepsilon)}(u), \tag{5.20}$$

where δ denotes the Dirac delta function, and $\chi_{[0,\varepsilon)}$ denotes the indicator function for the interval $[0, \varepsilon)$.

For criminals interactions

- Less experience criminals mimic the more experienced ones but also more experienced criminals, due to interactions, may increase their ability:

$$\mathscr{A}_{2,2}^2(u_* \to u | u_*, u^*) = \delta(u - (u_* + \beta_1 (1 - u_*) u^*)). \tag{5.21}$$

- Criminals chased by detectives are constrained to step back decreasing their activity value as the price to be paid for being caught. Furthermore, criminals are induced to return to the state of normal citizens with probability which increases with decreasing values of their level of criminality and increasing values of detectives skills:

$$\begin{cases} \mathscr{A}^1_{2,3}(u_* \to u|u_*, u^*) = \dfrac{1}{\varepsilon}\alpha_2(1-u_*)u^*\chi_{[0,\varepsilon)}(u), \\ \\ \mathscr{A}^2_{2,3}(u_* \to u|u_*, u^*) = (1-\alpha_2(1-u_*)u^*)\,\delta(u-(u_*-\gamma u^* u_*)). \end{cases} \tag{5.22}$$

- Criminals show a trend toward the mean ability of the lawbreakers environment:

$$\mathscr{B}^2_{2,2}(u_* \to u|u_*, \mathbb{E}^1_2) = \delta(u-(\beta_1 u_* + (1-\beta_1)\mathbb{E}^1_2)). \tag{5.23}$$

For detectives interactions

- Detectives gain experience by chasing criminals and increase their ability:

$$\mathscr{A}^3_{3,2}(u_* \to u|u_*, u^*) = \delta(u-(u_* + \gamma u^*(1-u_*))). \tag{5.24}$$

- Detectives show a trend toward the mean ability of the security forces environment:

$$\mathscr{B}^3_{3,3}(u_* \to u|u_*, \mathbb{E}^1_3) = \delta(u-(\beta_2 u_* + (1-\beta_2)\mathbb{E}^1_3)). \tag{5.25}$$

Model's parameters: The model includes the following seven parameters:

- $\eta^0 > 0$: Constant factor for microscopic interactions;
- $\nu^0 > 0$: Constant factor for microscopic–macroscopic interactions;
- $0 \le \alpha_1 < 1$: Susceptibility of citizens to become criminals;
- $0 \le \alpha_2 < 1$: Susceptibility of criminals to get back to normal citizens;
- $0 \le \beta_1 < 1$: Learning dynamics among criminals;
- $0 \le \beta_2 < 1$: Learning dynamics among detectives;
- $0 \le \gamma < 1$: Motivation/efficacy of security forces to catch criminals.

For additional technical details on the modeling approach, we refer the reader to [35]. This paper also presents a number of simulations, which show the role of parameters and initial conditions on the overall dynamics of the system. The next subsection will further enlighten, by a number of simulations, the predictive ability of the model.

5.4.4 Simulations and parameters sensitivity analysis

This subsection presents a number of simulations to enlighten both the influence of the initial conditions (not only mean value, but also shape of the probability distribution) and of the role of security forces over the dynamics. More in detail, the focus is on the following aspects:

1. Influence of the wealth distribution over the growth of criminality;
2. Role of the training of the security forces over the control of the criminality;
3. Influence of the mean wealth over the growth of criminality;
4. Role of the number of the detectives over the control of criminality.

Subsequently, the qualitative output of simulations will be summarized.

Dynamics for different shapes of wealth distribution: Simulations are developed corresponding to fixed values of the mean wealth, specifically two values are selected, low ($\mathbb{E}_1 = 0.2$) and high ($\mathbb{E}_1 = 0.6$), while two different shapes are considered for each case corresponding to higher and lower concentrations of wealth in the middle class, as depicted in Figs. 5.1a and 5.2a. In this case, we take a nonzero initial condition for the number of criminals, considering that $n_2(0)/n_1(0) = 0.05$ (corresponding to a society where the initial number of criminals is 5% of the number of citizens), and we define the quantity

$$\varphi(t) = \frac{1}{n_2(0)}(n_2(t) - n_2(0)) \times 10^2,$$

as a measure of the relative percentage change in the population of criminals. Simulations were developed for the set of parameters: $\alpha_1 = 0.0001$, $\alpha_2 = 0.15$, $\beta_1 = 0.1$, $\beta_2 = 0.9$ and $\gamma = 0.15$, while Figs. 5.1b and 5.2b report the evolution of φ corresponding to different wealth distributions.

It can be observed that a poorer society produces a growth in the number of criminals, that is still more accentuated for unequal wealth distributions. The model is capable to produce the opposite behavior for a richer society, giving a reduction in the number of criminals for the same choice of parameters.

Prevention of crime by training security forces: Consider a population initially distributed with $n_2(0)/n_1(0) = 0.05$ and with 500 detectives per 100,000 citizens and let us study the evolution of $\varphi(t)$ with variation of the parameters α_1 and γ. The initial distribution of detectives $f_3(0, \cdot)$ is a Gaussian-type function with mean value 0.5. Figure 5.3a shows the time dynamics of φ for different values of α_1, and for $\alpha_2 = 0.05$, $\beta_1 = 0.1$, $\beta_2 = 0.9$, and $\gamma = 0.5$. Figure 5.3b shows the same dynamics for $\alpha_1 = 0.0002$, and different values of γ. This figure confirms the empirical evidence that an effective action to fight crime consists in pursuing actions that contribute to reduce α_1 (education, employment, etc.,) and to improve the quality of citizens [192].

Dynamics for different mean wealth values: Simulations aim at depicting the time evolution of the number of criminals, starting from the ideal situation of $n_2(t = 0) = 0$, for different values of the mean wealth of citizens. With that purpose we consider two initial conditions for $f_1(0, u)$, which have a small rich cluster and a larger low-middle-class cluster with mean wealth taking values $\mathbb{E}_1 = 0.25, 0.53$. The solution is computed for large times, and Fig. 5.4a represents the final distribution of criminals

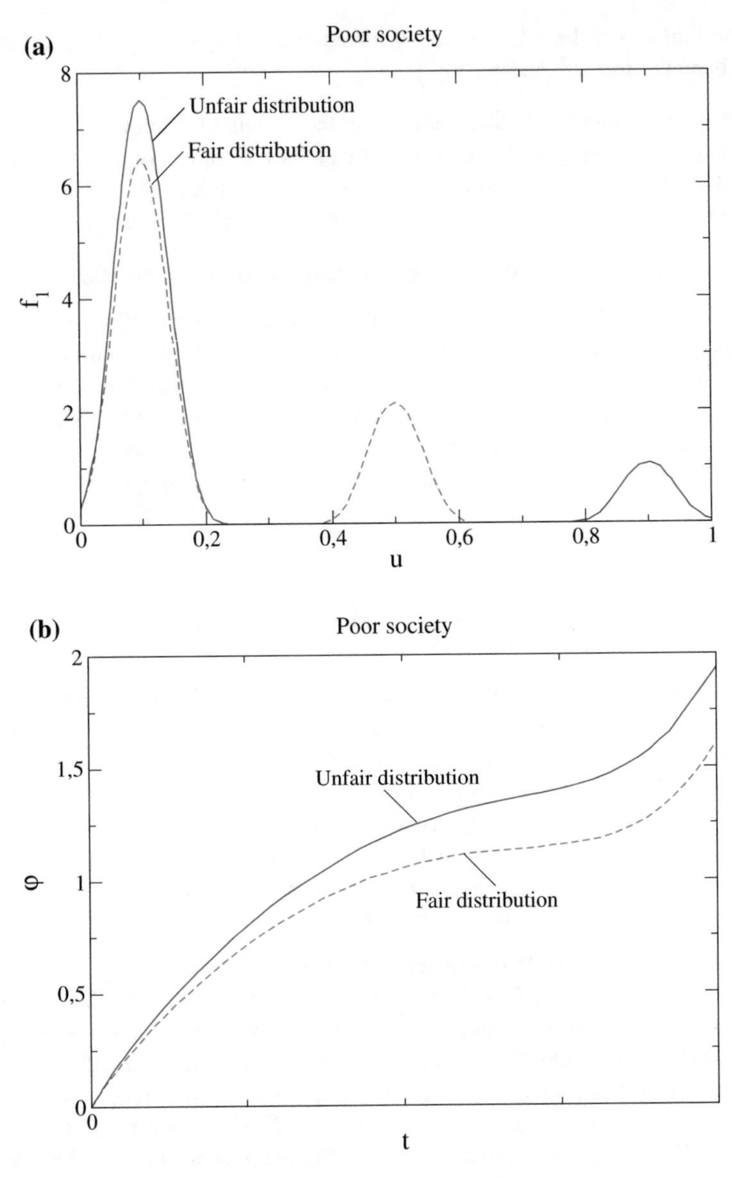

Fig. 5.1 (a) Initial wealth distributions for a poor society with $\mathbb{E}_1 = 0.2$. (b) Relative change of the population of criminals. $\alpha_1 = 0.0001$, $\alpha_2 = 0.15$, $\beta_1 = 0.1$, $\beta_2 = 0.9$, $\gamma = 0.15$

f_2. Figure 5.4b shows the evolution of the size of the population of criminals over time, $n_2(t)$, for these different values of $\mathbb{E}_1$. These figures show that a poor society leads to high levels of crime. This trend holds for a broad variety of parameters.

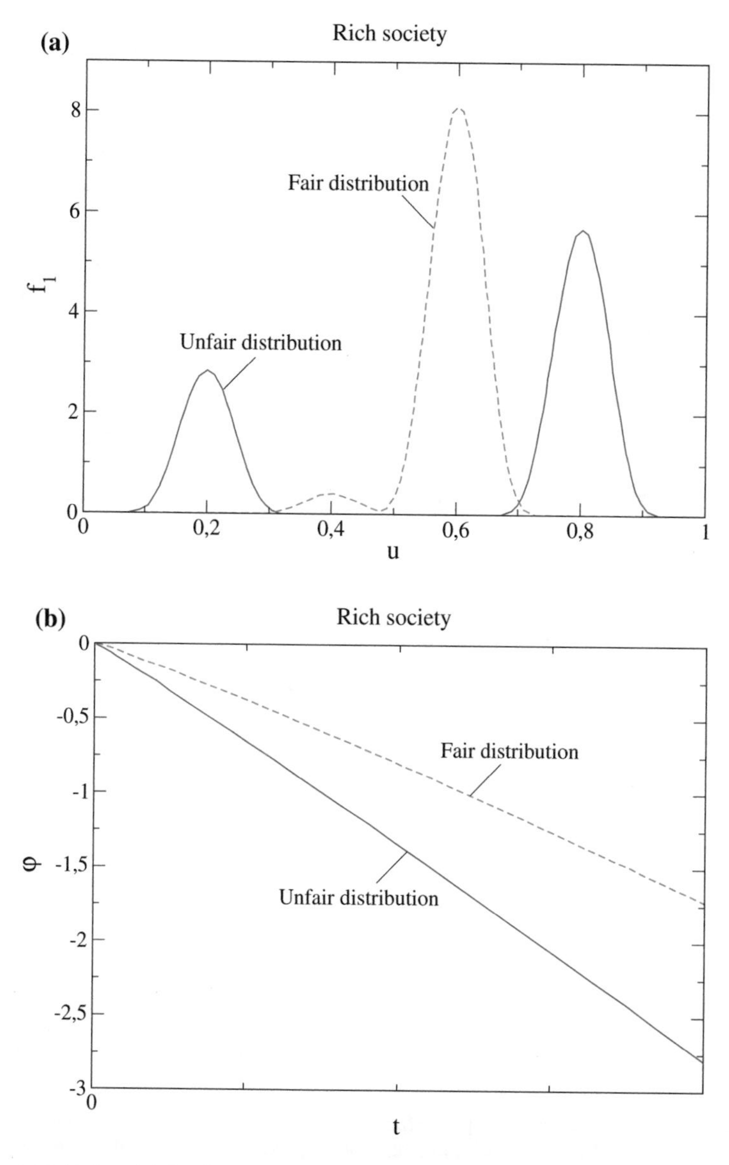

Fig. 5.2 (a) Initial wealth distributions for a rich society with $\mathbb{E}_1 = 0.6$. (b) Relative change of the population of criminals. $\alpha_1 = 0.0001$, $\alpha_2 = 0.15$, $\beta_1 = 0.1$, $\beta_2 = 0.9$, $\gamma = 0.15$

On the role of the number of detectives: Simulations investigate also the influence of the number of detectives in the development of crime. Taking into account the worldwide distribution of police agents per country, the rates have a median of 303.3 officers per 100,000 people and a mean of 341.8 officers [135]. Of course, these numbers depend on the social and structural differences between countries.

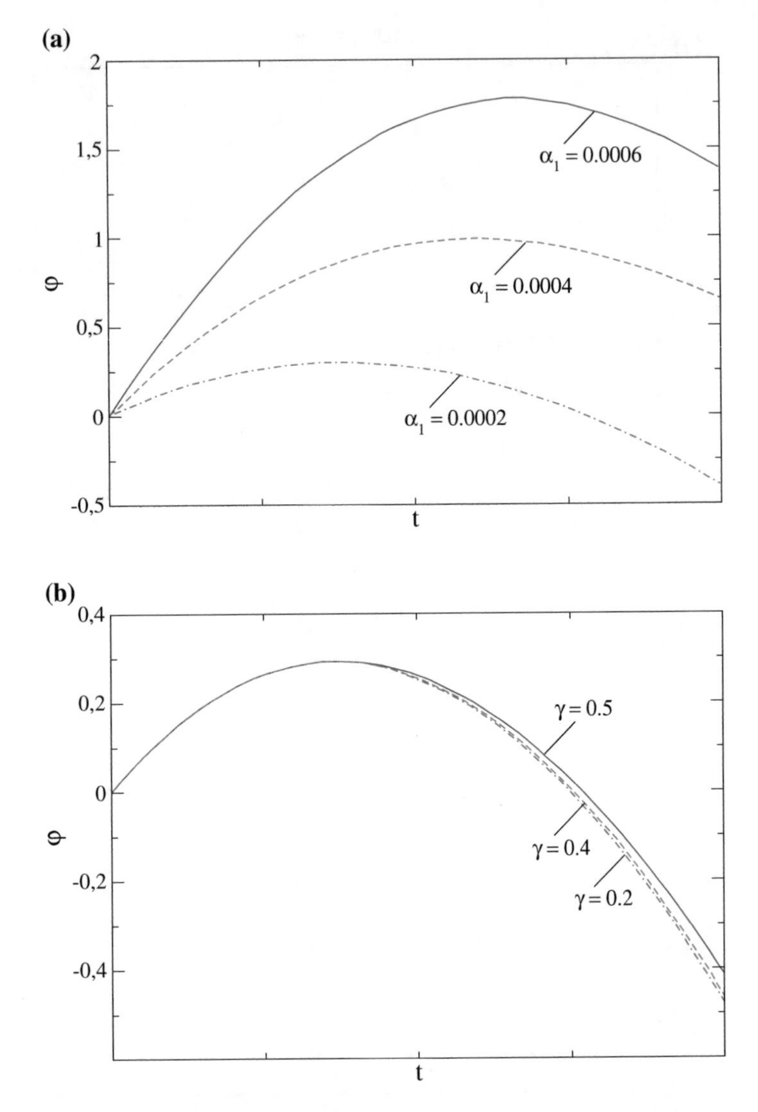

Fig. 5.3 (a) Relative change of the population of criminals for different values of citizens susceptibility to become criminals. (b) Relative change of the population of criminals for different values of motivation/efficacy of security forces to catch criminals. $\alpha_2 = 0.05$, $\beta_1 = 0.1$, $\beta_2 = 0.9$

Let us show the evolution of the number of criminals, taking a fixed initial mean wealth $\mathbb{E}_1 = 0.433$ and a fixed ratio $n_2(0)/n_1(0) = 0.1$, for different values of $n_3(0)/n_1(0)$ (in particular we consider two cases: 1110 and 330 detectives per 100,000 citizens). In all cases, the initial distribution of detectives $f_3(0, \cdot)$ is a Gaussian-type function with mean value 0.5.

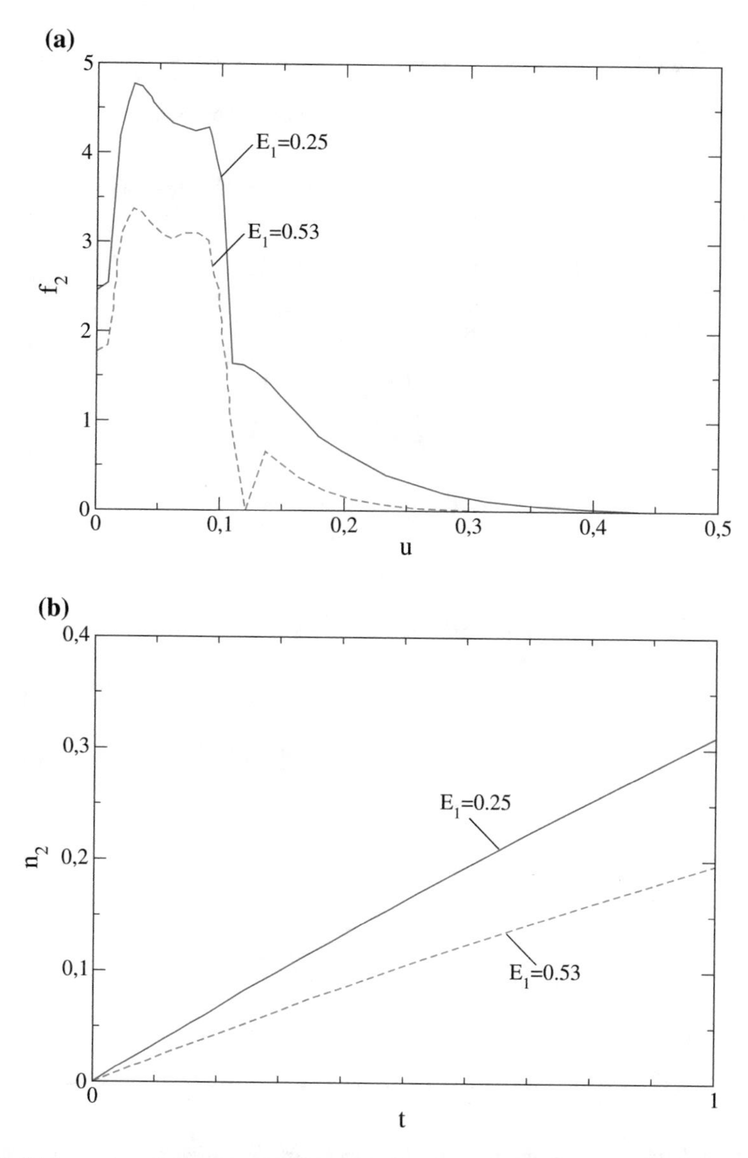

Fig. 5.4 (a) Large time distribution of criminals for two different mean wealth values. (b) Evolution of the size of functional subsystem 2, $n_2(t)$, for different values for the selected mean wealth values. $\alpha_1 = 0.01, \alpha_2 = 0.1, \beta_1 = 0.1, \beta_2 = 0.25, \gamma = 0.9$

Figure 5.5a shows the initial distribution of criminals $f_2(0, \cdot)$ and their large time distributions for different values of $n_3(0)/n_1(0)$. Figure 5.5b shows the evolution of the macroscopic quantities $\varphi(t)$ for different values of $n_3(0)/n_1(0)$. The results show that for this selection of parameters, the expected number cited in [135] keeps

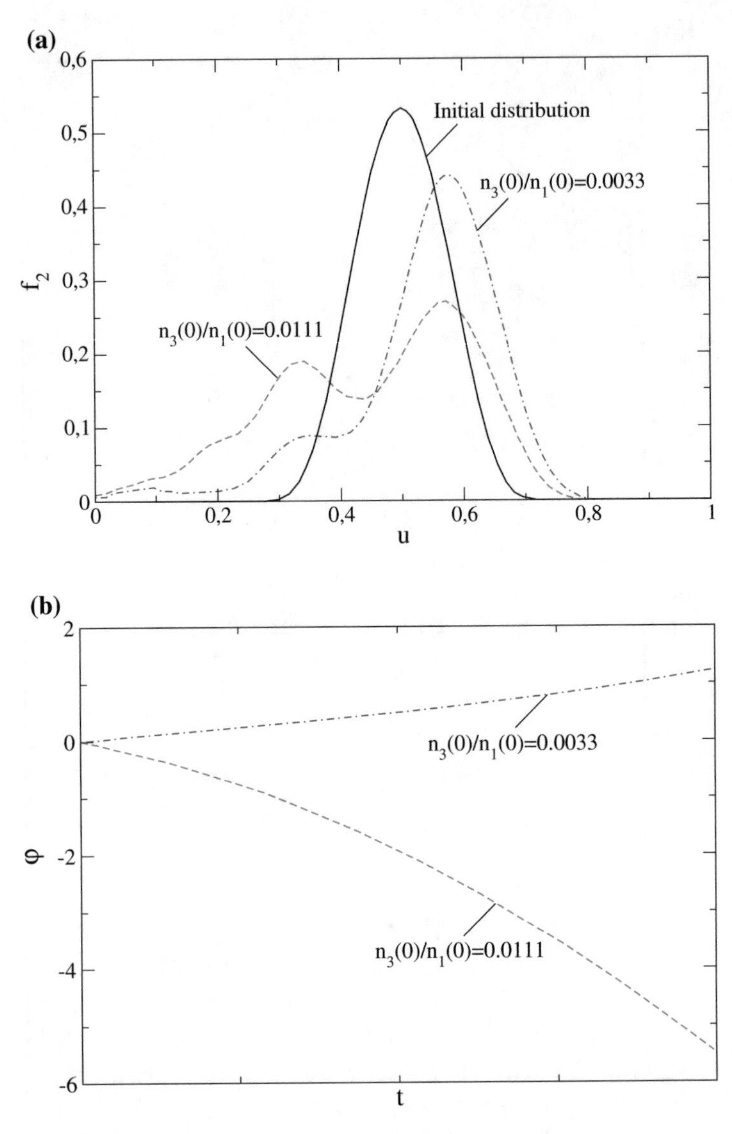

Fig. 5.5 (a) Initial and large time distributions of criminals, and (b) evolution of $\varphi(t)$. Continuous lines correspond to $n_3(0)/n_1(0) = 3.3 \cdot 10^{-3}$ and dashed lines correspond to $n_3(0)/n_1(0) = 11.1 \cdot 10^{-3}$. $\alpha_1 = 0.0001$, $\alpha_2 = 0.15$, $\beta_1 = 0.05$, $\beta_2 = 0.01$, $\lambda = 0.9$

the number of criminals under control. Larger squads contribute to the reduction in the number of criminals as well as in the mean criminal ability.

Summary of simulations results: The main results about the criminality dynamics which can be inferred from simulations are reported in Table 5.1.

Table 5.1 Criminality dynamics

Social context		*Poor society*	*Wealthy society*
Wealth distribution	equal	slow growth of criminals	slow decrease of criminals
	unequal	fast growth of criminals	fast decrease of criminals
Security forces training		not effective	effective
Number of detectives		effective	not effective
Mean wealth	high	low level of criminality	
	low	high level of criminality	

Summarizing, for what the influence of the *mean wealth* is concerned, for a variety of parameters, it results that the lower is the mean wealth of the society, the higher is the number of criminals and their criminal ability. Moreover, results are strongly affected by the *shape of wealth distribution.*

Indeed, a *fair distribution in a poor (wealthy) society* leads to a slow growth (decrease) of the number of criminals, while a fast growth (decrease) of their number is obtained if the wealth distribution is unfair. Low presence of middle classes sharpens the predictions of the model.

The *training of detectives* has an effective impact on the control of criminality in a wealthy society, while it is not influent in a poor society.

The *number of detectives* is important in a poor society, while the opposite occurs in a wealthy society. *Low presence of middle classes* enhances the aforementioned emerging prediction of the model.

5.5 Open Problems and Research Perspectives

This final section proposes some possible research perspectives which are brought to the attention of the interested reader looking ahead to future research programs. This section looks both at further developments of the systems approach to social sciences and at applications of the mathematical tools to the modeling of systems in different fields of life sciences. A list is proposed in the following being selected according to the authors' bias, however, without claim of completeness. In more detail, the said perspectives focus on the following topics:

1. Further developments on the modeling the dynamics of social systems;
2. Modeling immune competition including proliferative/destructive dynamics as well as mutations and selection;
3. Revisiting the classical population dynamics with the aim of including specific complexity features of living systems;
4. Analytic problems.

These topics are presented in the next subsections at a qualitative level leaving to the interested reader their development within possible research programs.

5.5.1 Perspectives in modeling social dynamics

This chapter has been focused on the modeling and simulation of social systems. Both the theoretical approach and the applications leave a variety of open problems that are, in our opinion, worth to be developed. These open problems can be mainly focused on modeling issues to be properly developed to allow models to depict complex scenarios not yet treated in the literature. A number of open problems are proposed in the following, where the presentation of each problem is followed by some remarks, "in italic," to report what has been already done in the literature.

• **Dynamics of cooperation and competition**: The separation between these two dynamics is modeled by a threshold, which separates consensus from dissent. According to the approach of this chapter, the threshold depends on the distance between the interacting elements.

In some cases, for example, opinion formation, small distances induce consensus, while the opposite trend is observed in the case of social competition. It is worth mentioning that such a threshold cannot be considered a constant quantity, as it depends not only on external actions, but also to the overall state of the system.

More in general both types of dynamics can occur independently from the aforementioned drastic separation, but one can be dominant with respect to the other depending on Governments' actions.

Understanding the specific features that promote consensus has been a constant subject of research activity by mathematicians and physicists; examples are given by [145, 177]. *The recent survey* [177] *provides a deep analysis on this topic, where the dynamics toward consensus (cooperation) is linked to the aforementioned threshold. The trend is increased (or decreased) according to the dynamics and external actions, which modify such a threshold. Papers* [46, 107] *present specific applications related to this topic. These papers have already developed some aspects of this topic and discovered that if applied in practical cases it can predict events that cannot be easily predicted.*

• **Dynamics over networks**: This chapter devoted a certain emphasis to this topic appears to be very important in certain fields of social sciences, for example, opinion formation, where the structure of the network can contribute to the diffusion of opinions. It is worth stressing that in the approach proposed in this book, the partition into functional subsystems is not organized according to universal rules, but it is related on the specific investigation which is pursued. As a consequence, the organization into a network follows the same hallmarks. In addition, it is not fixed for all times as the dynamics of subsystems might induce changes in the network to be viewed as a dynamical system. The formation of networks is one of the challenging research perspectives indicated in [68], where the structure of the network is allowed to evolve in time due to interactions.

The book [68] *brings to the reader's attention an important problem, namely the need of understanding how the structure of a network is generated by interactions. Arguably, this complex dynamics can be related to a deep understanding of interactions and collective learning dynamics recently developed in* [75] *based on the idea that understanding the dynamics can contribute to modeling interactions. According to* [75], *collective learning is modeled by a combination of affinity estimates and subsequent learning processes. A possible conjecture is that the structure of a network is a consequence of affinity, namely nodes are created by groups with closed affinity. Indeed, it is a dynamical process as learning modifies affinity and hence the said structure.*

• **Modeling interactions**: A deep understanding of the dynamics of interactions can provide an important contribution to the study of collective dynamics. The mathematical structure proposed in this chapter can transfer the individual dynamics into collective behaviors. Empirical data should be organized not only toward modeling interactions, but also to understand how theoretical contributions from social sciences can provide a theoretical approach to the aforementioned modeling. An additional perspective refers to interactions involving multiple dynamics.

Some recent papers see, for example [46, 107], *have shown that the interactions of different dynamics can explain social phenomena which are not foreseen when one only dynamics is taken into account. In general, interactions can induce in some cases mutations and Darwinian selection, where some social systems, for instance political parties, are characterized by a critical (small) size below which interactions do not transfer anymore individuals to a different functional subsystem. Likewise, it might exist a* critical (large) *size such that if such threshold is overcome, individuals are induced to move into another functional subsystem. In principle, this remark applies to all types of group of interest.*

Three topics have brought to the reader's attention after a selection, according to our bias, out of several possible ones. Of course additional topics can be selected. All possible perspectives generate challenging analytic problems that will be treated in the last subsection.

5.5.2 *Modeling multicellular systems and immune competition*

A well-defined application has been presented and critically analyzed in the preceding section of this chapter. However, the mathematical structures of Chapter 3 can be applied in a variety of fields in life sciences. An interesting field of application is the modeling of the immune competition in the presence of mutations and selection. The dynamics includes also proliferative and destructive events. In addition, interactions can generate mutations into new populations, eventually followed by Darwinian-type selection. Therefore, mathematical structure more general than that used for the modeling of social dynamics is needed. In particular, conservation of number of particles cannot be any more applied, and the number of functional subsystems evolves in time.

A brief overview of the existing literature on this type of approach is proposed before dealing with technical issues and, in particular, to define the properties of the structure to be used for modeling. The contents of this subsection take advantages of some recent papers, more precisely [57], where Darwinian dynamics was introduced as a substantial improvement of previous studies on the immune competition [58], as well as of [91], and [40], where further developments of [57] are proposed toward the development of a new systems approach to the modeling of biological systems.

Let us now focus more precisely to large systems of interacting cells. As it is known [237, 242], the onset and development of genetic diseases has an origin related to the dynamics at a cellular level as it can contribute to a dynamics at a cellular level when, during replication, the daughter cell exhibits a phenotype different from that of the mother cell, despite both belong to the same genotype [73]. This dynamics can be viewed within the general theory of evolution [174]. Paper [91] indicates some features of the dynamics which should be retained by mathematical models. Quoting [91]:

> **Interactions include proliferative events, which normally generate individuals with the same phenotype, but also, although with a small probability, new individuals, with a different phenotype: either more or less suited to a changing environment. As a consequence, the number of populations can evolve in time.**
>
> **If a number of populations constitutes the inner system that interacts with the external environment that evolves in time due to the aforementioned interaction. The distribution of the phenotypes is heterogeneous and can evolve in time. The net proliferation rate of the newborn population depends on its fitness with respect to the environment.**
>
> **The dynamics can result in with the extinction of the less suitable populations and survival of the most suitable ones. In some cases more than one population can survive.**

Focusing on mutations and selection at the scale of cells, studied in the already cited paper [91], the dynamics is such that the mother cell can generate, in the reproduction process, a cell mutated with respect to her specific features. This newborn cell might be more well fitted (or less well fitted) with respect to the environment and then reproduce with a rate greater than that of the mother cell and invade the tissue (or reproduce with a rate lower than that of the mother cell and then progressively disappear).

Therefore, the environment plays an important role as newborn individuals, namely offspring, more well fitted to the environment survive and proliferate, while others which are less fitted may go to extinction. In general, the same genotype can generate more than one phenotype populations, which survive in a certain environment.

Let us stress that paper [91] provides a general mathematical structure suitable to include a variety of specific biological dynamics such as the immune competition in the presence of mutations and selection [57]. Moreover, it contributes to the challenging objective of reaching a deeper understanding of the laws of Darwinian evolutionary theory.

An additional important feature to be taken into account is the multiscale characteristics of all biological systems. In fact, if we agree that the scale of cells is the *microscale*, then the dynamics at this specific scale is induced, as it is known, by the dynamics at the lower *molecular scale* corresponding to genes in the DNA structure [115, 237]. Letting the distance between cells to zero, in some appropriate limit, allows to derive the equations modeling tissues as shown in [33]. Therefore, the structure of tissues can evolve in time as the properties of cells depend on the dynamics at the lower molecular scale. A network of tissues can constitute organs whose functions also evolve in time. The book by Frank [115] is mainly devoted to understand the link between the molecular and the cellular scale.

The search for mathematical structures to model multiagent biological systems initiated with the pioneer paper [41] focusing on the immune competition between cancer and immune cells. A systematic analysis of the aforementioned classes of models is given in the book [58], where the whole path, phenomenological analysis of the biological system, modeling, qualitative analysis, simulations, and derivation of macro-scale models, is delivered in a unified presentation.

An important development is given in [92, 96] concerning the dynamics of a virus undergoing successive mutations somehow contrasted by the immune system. The merit of these papers consists in having introduced the concept of Darwinian-like mutations and having modified accordingly the known structures.

The hint given by aforementioned papers was taken in [57] to model the evolutive dynamics of cancer cells related to replication dynamics, and the corresponding evolution of the immune system [81, 134] through a complex learning dynamics. A structure suitable to include interactions with external environments is proposed in [91].

Let us now focus on the features of the structure to be used for the modeling approach and on the related analytic and computational difficulties. As mentioned, the interaction operator includes proliferative and destructive events. Therefore, the number of active particles is not constant in time. In addition, interactions with the external environment need to be modeled. However, the most complicated feature is the variable number of equations, which can grow or decrease due to mutations that can generate new functional subsystems, e.g., cell populations, that contrasted by the immune system can become either invasive due to their higher proliferative ability or go to extinction due to a weaker ability to contrast the immune system.

This complex dynamics has been studied in [57], where a general differential framework has been proposed according to the conjecture proposed by a team of immunologists [81], who argued that the immune system evolves by mutations following the mutations of cancer cells. The authors have also shown that for certain choice of the parameters, the system can even evolve toward an asymptotic equilibrium, where the number of cancer cells reaches a constant value under the action of the immune system that does not, however, succeed in depleting the tumor cells.

This is a very special case of a dynamics depicted by the model which, however, has not been fully analyzed. Hopefully, additional analysis will enlighten the properties of the model we are discussing about. More in details we conjecture that the equilibrium configurations found in [57] are rare critical states that separate the blowup of cancer cells of the last mutated populations and the reaction of the immune system which might succeed to suppress cancer cells.

The model in [57] includes the transition of epithelial cells into the first population of cells with proliferative competence. It would be interesting refining the assumption of constant transition rate which can generate a decay apparently successful, but followed by a new competition with growth and decay. As an example, one might investigate the laps of time during which the dynamics of transition into the first populations occurs so that both the rate of transition and the time interval during which such transition occurs are studied.

In other words, the approach proposed in [57] deserves further developments that can hopefully take advantage of the mathematical structure proposed in this chapter. Of course, the approach should take into account that the lack of number conservation and the variable number of equations induces new challenging analytic and computational problems.

5.5.3 *Revisiting population dynamics*

Models of population dynamics provide a description of the dynamics of living systems by defining the time evolution of the number of individuals for each of interacting populations. The book by Perthame [199] provides a nice presentation of population models showing how the literature developed from the original Lotka–Volterra approach to models that account for delay terms, internal variables [240], and various other features. This book also presents the main concepts related to the qualitative analysis of problems and applications in biology. Additional useful bibliography can be obtained in the books by Bürger [72], by Diekmann et al. [104, 105], and by Thieme [232].

The advantage of population dynamics is that computational problems are less expensive and that the qualitative analysis provides, in several cases, the needed information on the asymptotic behavior of the solutions, while this behavior can be obtained, in the case of kinetic models, only by simulations. On the other hand various important features, that are typical of complex systems, are not included in the simple description of population dynamics.

Therefore, the revising should be addressed to insert at least a part of these features in models of population dynamics. For instance, hybrid models might be studied, where in a dynamical system suitable to describe the interaction of various functional subsystems, some of them are deterministic, namely simply described by the number of individuals, and other are modeled by the kinetic theory of active particles.

The basic idea, already introduced in [79], is that each populations is linked to an internal variable, corresponding to the activity, whose dynamics is determined by interactions. Then, the model is expressed by a system of differential equations with random parameters evolving in time.

A recent contribution [127] proposes an approach, where deterministic populations are coupled to stochastic populations. The authors derive an equation to model the time evolution of an entropy function to depict how stochastic behaviors develop in time. Hence the dynamics couples heterogeneous populations, represented by stochastic, and homogeneous ones represented by deterministic variables. The study of the entropy functions leads to a quantitative analysis of the role of heterogeneity on the overall dynamics of the system.

5.5.4 Analytic problems

Analytic problems are generated by the qualitative analysis of the solutions of the initial value problems generated by the application of models to the study of real-world systems. Local existence and uniqueness of solutions can be obtained by almost straightforward applications of fixed-point theorems [203]. In fact, the structure of the equations guarantees, under appropriate Lipschitz continuity assumptions of the integral operators, local existence. Moreover, if the number of particles is preserved, the solutions can be prolonged for arbitrary long times as the aforementioned structure assure preservation of the L_1 norm. This feature is rapidly shown in paper [14], where the proof has been given for closed systems, while the technical generalization to open system should still be developed.

On the other hand, when the said number is allowed to grow, as it happens in the applications described in Subsection 5.5.2, this type of analysis requires far more sophisticated tools, namely sharp functional inequalities. This explains why existence analysis has been developed only for some special models such as the one treated in [57].

The most challenging problem remains the analysis of asymptotic behaviors for systems where simulations show a trend to equilibrium solutions. The application of Schauder fixed-point theorem shows existence at least of one equilibrium solution. However, uniqueness and stability are not proved. A technical difficulty, not yet solved, is that the shape of the asymptotic configuration (shape of the probability distribution) depends on the shape of the initial datum. This problem appears to be a very challenging target for applied mathematicians. Indeed, it is important for the applications as models are required to predict the asymptotic trend of social dynamics so that acceptable behaviors can be strengthened, while not acceptable behaviors can be weakened.

Finally, let us mention that the various models proposed in this chapter have been derived in the case of space homogeneity. If a space structure is included in the model, then new challenging problems are generated to constitute challenging objectives for applied mathematicians.

Chapter 6
Mathematical Models of Crowd Dynamics in Complex Venues

6.1 Plan of the Chapter

This chapter is devoted to the derivation of mathematical models of large systems of active particles which move and interact in space. The content focuses mainly on crowd dynamics; however, the dynamics of multicellular systems also treated and some perspective ideas are given on vehicular traffic and swarms. The modeling approach is based on the use and developments of the mathematical structures proposed in Chapter 3.

The literature on the modeling of pedestrian crowds has rapidly developed in the last decade due to the potential benefits that the study of these systems can bring to the society, for instance to care about safety in evacuation dynamics [209–211, 245]. In addition, applied mathematicians have been attracted by the challenging analytic and computational problems generated by the derivation of models and by their application to real-world dynamics.

The plan of the chapter is as follows:

- Section 6.2 presents a general introduction and critical analysis of the literature on crowd dynamics and motivates the modeling approach treated in this chapter. More in detail, the concept of scaling and representation of the system at the microscopic, macroscopic, and mesoscopic scales are discussed. Furthermore, mathematical structures are presented which are deemed to offer the conceptual basis for the derivation of models at each scale.
- Section 6.3 defines the conceptual hallmarks which guide the application of the tools of Chapter 3 to the modeling and analysis of crowd dynamics. The concept of *behavioral social dynamics* is introduced as a new approach to dynamics, where interactions depend on the behavior of the interacting entities viewed as active particles. Subsequently, a mathematical structure is derived to provide the conceptual basis for the modeling approach at the mesoscale.

N. Bellomo et al., *A Quest Towards a Mathematical Theory of Living Systems*, Modeling and Simulation in Science, Engineering and Technology,
DOI 10.1007/978-3-319-57436-3_6

- Section 6.4 takes advantage of the framework presented in the preceding section to derive models which include also the dynamics of individuals which change the rules of their participation to the dynamics. Models in unbounded and bounded domains are presented, and a discussion of the modeling of stressful conditions, is developed.
- Section 6.5 presents some sample simulations with a special focus on evacuation dynamics from venues with complex geometry. These simulations, which are obtained by the Monte Carlo particle methods reviewed in Chapter 2, enlighten the role of stressful conditions on the evacuation dynamics. The problem of the model's validation is also discussed.

- Section 6.6 shows how some models can be developed by an approach different from that used for crowd dynamics. Namely, it is shown how macroscopic equations can be derived from the underlying description delivered by the kinetic theory and subsequently simulations are developed by classical deterministic methods.

- Section 6.7 concludes this chapter by looking ahead to some possible research perspectives. The presentation focuses on vehicular traffic and swarm dynamics. Some analytic and computational problems are also presented. These topics include social dynamics generated by crowd heterogeneity, topological interactions, and multiscale problems.

6.2 Overview on Crowd Modeling

The literature on crowd modeling is reported in some survey papers, which offer to applied mathematicians different viewpoints and modeling strategies in a field where a unified, commonly shared, approach does not exist yet.

The review paper [139] introduces the main features of the physics of crowd viewed as a multiparticle system and focuses on the modeling at the microscopic scale for pedestrians undergoing individual-based interactions.

The book [85] essentially deals with the modeling at the macroscopic scale, by methods analogous to those of hydrodynamics, where one of the most challenging conceptual difficulties consists in understanding how the crowd, viewed as a continuum, selects the local speed and velocity direction [157].

The survey [50] introduces the concept of the crowd as a living, hence complex, system and subsequently the search of mathematical tools suitable to take into account, as far as it is possible, their complexity features.

The hints proposed in [50] have been developed in [29] to model the dynamics in unbounded domains, and, subsequently, in [43] for a crowd in domains with boundaries and internal obstacles. These two papers introduce the concept of *behavioral crowd*, namely a crowd whose dynamics depends on the strategy and behaviors that pedestrians develop based on mechanical ad social interactions with the other pedestrians. The modeling of such system requires new ideas and mathematical tools, and this challenge has motivated the growing interest of applied mathematicians.

This section provides the conceptual basis for the modeling approach proposed in the next section. More in detail, the main features of the crowd viewed as a behavioral system are assessed, and the mathematical structures for the derivation of models at the microscopic, mesoscopic, and macroscopic scales are briefly reviewed.

6.2.1 Features of Behavioral Crowds

Referring to [142], the following definition of crowd is adopted:

Definition of crowd: Agglomeration of many people in the same area at the same time. The density of people is assumed to be high enough to cause continuous interactions, or reactions, with other individuals.

The dynamics of a crowd cannot simply rely on mechanical and deterministic causality principles. Indeed, the heterogeneous behaviors of pedestrians and the social dynamics related to their communications can have an important influence over their strategy.

According to [43], whatever modeling approach is employed, the following complexity features should be properly taken into account:

- **Ability to express a strategy:** Pedestrians are capable to develop specific strategies, which depend on their own state and on that of the individuals in their surrounding environment. The strategy depends also on the venue where walkers move.
- **Heterogeneity and hierarchy:** The ability to express a strategy is heterogeneously distributed due to different walking abilities. Heterogeneity can also refer to social behaviors, such as the aggressiveness in a crowd where two groups contrast each other or panicking behaviors where the ability to pursue optimal strategies is typically lost. Furthermore in crowds there may be leaders who aim at driving all other pedestrians to their own strategy.
- **Nonlinear and non-local interactions:** Walkers' strategy depends both on mechanical and social interactions. Interactions are nonlinearly additive and involve immediate neighbors, but also distant individuals. Interactions occur both between pedestrians and between a pedestrian and the environment where he moves. The latter are affected by different geometrical and environmental features, such as abrupt changes of directions, luminosity conditions, and many others.

These complexity features are a specialization to the crowd dynamics of the general ones we have discussed in Chapter 1. The choice of kinetic theory methods is motivated by the need of a mathematical approach consistent with these features.

6.2.2 Scaling, Representation, and Mathematical Structures

Models can be developed at the following three classical scales:

- *Microscale*: Pedestrians are individually identified, and the state of the system is delivered by the whole set of their positions and velocities. Mathematical models are generally stated in terms of large systems of ordinary differential equations, where the dependent variables model position and velocity.
- *Mesoscale*: The microscopic state of the interacting individuals is still identified by the position and velocity, but their representation is delivered by a suitable probability distribution over the microscopic state. Mathematical models describe the evolution of the above distribution function generally by nonlinear integro-differential equations.
- *Macroscale*: The system is described by gross quantities, namely density and linear momentum, regarded as dependent variables of time and space. These quantities are obtained by local average of the pedestrians' microscopic state. Mathematical models describe the evolution of the above variables by systems of partial differential equations.

It is worth noticing that these three scales are closely related. As discussed into more depth in the following, the mesoscopic approach needs to model the dynamics at the microscopic scale. In turn, suitable asymptotic methods and limits applied to mesoscopic models lead to the derivation of models at the macroscopic scale. This topic has been dealt with in a recent paper [27] which develops previous studies on vehicular traffic [32].

Let us consider the representation of a crowd at the different scales and the related mathematical structures. This overview contributes to a deeper understanding of the kinetic theory approach presented in the next sections. In the following, Σ symbolically represents the overall geometrical features of the venue where pedestrians move, and $\alpha \in [0, 1]$ is a parameter accounting for the quality of the venue, from the worse quality, $\alpha = 0$, which prevents the motion, to the best quality, $\alpha = 1$, which allows the fastest motion.

- *Microscale* : The microscopic description is provided by the *position* $\mathbf{x}_i = \mathbf{x}_i(t)$ and the *velocity* $\mathbf{v}_i = \mathbf{v}_i(t)$ of each walker, with $i \in \{1, \dots, N\}$. Coordinates of the position of each walker are made dimensionless by division with a characteristic length ℓ. Likewise, the components of the velocity are divided by the limit speed v_ℓ corresponding to that of a fast walker moving in a high-quality venue.

The mathematical structure consists in a set of $2 \times N$ ordinary vector differential equations:

$$\begin{cases} \dfrac{d\mathbf{x}_i}{dt} = \mathbf{v}_i \,, \\[2ex] \dfrac{d\mathbf{v}_i}{dt} = \mathbf{F}_i(\mathbf{x}_1, \dots, \mathbf{x}_N, \mathbf{v}_1, \dots, \mathbf{v}_N; \alpha, \Sigma). \end{cases} \tag{6.1}$$

The key term of the modeling is $\mathbf{F}_i(\cdot)$ corresponding to the psycho-mechanical acceleration of the i-th walker based on the action of other walkers in its influence domain. Empirical data [179, 217] can contribute to modeling the aforementioned acceleration at the microscopic scale. The main contributions to this approach have been given by Helbing and coworkers [139, 142, 143].

• *Macroscale* : The macroscopic description is provided by the local *density* $\rho = \rho(t, \mathbf{x})$ and the *mean velocity* $\mathbf{V} = \mathbf{V}(t, \mathbf{x})$.

The mathematical structure defines conservation of mass and dynamical equilibrium of linear momentum:

$$\begin{cases} \partial_t \rho + \nabla_{\mathbf{x}} \cdot (\rho\, \mathbf{V}) = 0\,, \\[2ex] \partial_t\, \mathbf{V} + (\,\mathbf{V} \cdot \nabla_{\mathbf{x}})\, \mathbf{V} = \mathbf{A}[\rho, \mathbf{V}; \alpha, \Sigma], \end{cases} \tag{6.2}$$

where $\mathbf{A}[\rho, \mathbf{V}; \alpha, \Sigma]$ is a psycho-mechanical mean acceleration acting on walkers in the elementary space domain $[\mathbf{x}, \mathbf{x} + d\mathbf{x}] = [x, x + dx] \times [y, y + dy]$. This acceleration depends on α and Σ as in the microscopic case even though it is now a locally averaged quantity. Square brackets are used to denote functional, rather than function, maps. A typical example is when dependence on gradients is considered.

Models at the macroscopic scale can also be based only on the mass conservation [157] according to the following structure:

$$\begin{cases} \partial_t \rho + \nabla_{\mathbf{x}} \cdot (\rho\, \mathbf{V}) = 0\,, \\[2ex] \mathbf{V} = \mathbf{V}[\rho; \alpha, \Sigma]\,, \end{cases} \tag{6.3}$$

where the second equation is delivered by a phenomenological interpretation of the strategy of walkers and links the mean velocity to local density conditions, depending on α and Σ.

Derivation of models at the macroscopic scale has been conceptually introduced by Henderson [148] and subsequently transferred to a framework with some analogy to fluid dynamics by Hughes [156, 157]. These papers provide a method to compute walkers trajectories corresponding to a criterion which optimizes the search of the exit. The criterion has been further refined based on optimal transport theory. With a few exceptions [38], first-order models have been used. The book [85] provides an exhaustive presentation of macroscopic type models, including those derived by conservation of probability measures.

Experimental activity toward the collection of empirical data provides several valuable contributions (without claim of completeness, see [179, 180, 216, 217]). An open problem is the modeling of the strategy by which pedestrians organize their movement in stressful conditions.

• *Mesoscale* : The mesoscopic description is provided by the *probability distribution function* over the microscopic state of pedestrians $f = f(t, \mathbf{x}, \mathbf{v})$ defined such that

$f(t, \mathbf{x}, \mathbf{v})\, d\mathbf{x}\, d\mathbf{v}$ gives the number of individuals who, at time t, are located in $[\mathbf{x}, \mathbf{x} + d\mathbf{x}]$ with velocity in $[\mathbf{v}, \mathbf{v} + d\mathbf{v}]$.

As already shown in Chapters 2 and 3, macroscopic quantities are obtained by velocity-weighted moments. As an example, local density and flux are obtained as follows:

$$\rho[f](t, \mathbf{x}) = \int_{D_{\mathbf{v}}} f(t, \mathbf{x}, \mathbf{v})\, d\mathbf{v}, \tag{6.4}$$

and

$$\mathbf{q}[f](t, \mathbf{x}) = \int_{D_{\mathbf{v}}} \mathbf{v}\, f(t, \mathbf{x}, \mathbf{v})\, d\mathbf{v}, \tag{6.5}$$

where $D_{\mathbf{v}}$ is the domain of the velocity variable, while the mean velocity is

$$\xi[f](t, \mathbf{x}) = \frac{\mathbf{q}[f](t, \mathbf{x})}{\rho[f](t, \mathbf{x})}. \tag{6.6}$$

The dynamics is obtained by a balance of microscopic entities in the elementary volume of the space of microscopic state. This amounts to equate the transport of f to the net flow due to interactions. The formal result is as follows:

$$(\partial_t + \mathbf{v} \cdot \nabla_{\mathbf{x}})\, f(t, \mathbf{x}, \mathbf{v}) = \big(J^+ - J^-\big)[f; \alpha, \Sigma](t, \mathbf{x}, \mathbf{v}), \tag{6.7}$$

where J^+ and J^- model the inlet and outlet fluxes induced by interaction among pedestrians and between pedestrians and walls. More details on these terms are given in the next section, where the role of the activity variable is enlightened.

The literature on modeling human crowds by kinetic theory methods is far less developed than the one at the other scales. However, various innovative contributions have been proposed in recent years. The hints proposed in [39, 50, 155] have been developed in [29] for a dynamics in unbounded domains and in [43] to include interaction with boundaries.

As previously mentioned, these two papers introduce the concept of *behavioral crowd*, namely of a crowd whose dynamics strongly depends on the strategy and behaviors that pedestrians develop based on mechanical and social interactions with the surrounding pedestrians. Validation of models has been treated in [44]. A hierarchy of models is studied in [94], which provides an important conceptual framework for further developments.

Remark 6.1 *It is useful, in the modeling approach, using dimensionless quantities. For example, linear space variables x and y are referred to the largest dimension ℓ of the venue; number density is referred to the maximal density corresponding to the largest admissible density ρ_M corresponding to packing of walkers; speed, namely the velocity modulus, is referred to v_L corresponding to the highest admissible velocity of a walker in a high-quality venue. According to this normalization, all practical quantities take value in the interval* $[0, 1]$.

Remark 6.2 *The individual or collective dynamics, at all scales, should take into account also the psycho-mechanical to model the strategy that each walker develops, including possible stress conditions which appear induced by fear due to lack of safety. Such a strategy depends not only on the interactions between walkers, but also on the quality and on the overall shape of the venue, namely on* α *and* Σ.

Finally, it is worth stressing that the present state of the art only partially achieves the challenging objective described in the above remark. The model presented in the next two sections shows a certain ability toward a partial achievement. Subsequently, possible research perspectives are presented with the aim of improving the descriptive ability of models.

6.3 On the Kinetic Theory Approach to Behavioral Dynamics

As already mentioned, the modeling approach of a heterogeneous crowd cannot rely on the deterministic causality principle typical of classical mechanics since the strategy developed by a pedestrian depends on his individual interpretation of those of the others. Such an observation is at the basis of the $behavioral social dynamics$. This section anticipates some terminology and some preliminary ideas of the approach that will be developed in the following.

- The modeling approach is based on suitable developments of the so-called kinetic theory for active particles, which applies to systems composed of many interacting entities [47]. Hence the $mesoscale$ representation is chosen.
- Pedestrians, namely the $microscopic system$, are viewed as $active\ particles$, that have the ability of expressing their own strategy, called $activity$. This ability can differ for different individuals in the same crowd, being understood that the activity is heterogeneously distributed.
- The overall system is subdivided into groups of pedestrians who share common features according to the hallmarks of a systems theory of social systems introduced in [4] and followed in [46, 107]. We refer to these groups as $functional subsystems$ (FSs). The subdivision of the crowd in FSs permits to account for different walking abilities, strategies and/or the presence of leaders.
- Pedestrians can move across functional subsystems due to a $social dynamics$ generated by communications. Therefore, both strategy and dynamical rules followed in the movement may change in time.
- Each functional subsystem is described by a $probability distribution$ over the microscopic state of pedestrians, while interactions are modeled by theoretical tools of game theory [186].

Applications of this approach to the modeling of crowds and social systems are proposed in [27, 29]. In the next subsections, a more detail discussion is provided about the representation of the system, the modeling of microscopic scale interactions, and the derivation of a mathematical structure consistent with the complexity paradigms of behavioral dynamics.

6.3.1 Mesoscopic Representation of a Crowd

Let us now consider a crowd which is subdivided into different functional subsystems. The microscopic state of pedestrians, viewed as active particles, is defined by position $\mathbf{x}$, velocity $\mathbf{v}$, and activity u. Since the dynamics is in two space dimensions, polar coordinates are used for the velocity variable, namely $\mathbf{v} = \{v, \theta\}$, where v is the speed and θ denotes the velocity direction.

The *mesoscopic (kinetic) representation* of the overall system is delivered by the probability distribution at time t, over the pedestrians microscopic state:

$$f_i = f_i(t, \mathbf{x}, v, \theta, u), \quad \mathbf{x} \in \Sigma, \quad v \in [0, 1], \quad \theta \in [0, 2\pi) \quad u \in [0, 1], \tag{6.8}$$

for each functional subsystem labeled by $i = 1, \ldots, n$.

Remark 6.3 *Let us notice that the activity variable u has been introduced to account for some emotional state of the walker. The specific assessment of this variable is also related to the specific investigation of the model. As an example, which will be extensively treated later, this variable can model, in evacuation dynamics, the stress of walkers. This variable can substantially modify the walking strategy, for instance, by increasing the speed and enhancing clustering phenomena. An additional comment is that the subdivision into functional subsystems can be induced by different features, for instance, walkers with different abilities or walking toward different targets.*

Calculations of macroscopic quantities can be developed as we have already seen in the preceding section. Technical calculations should, however, account for the introduction of the activity variable and of the use of polar coordinates that are useful, as we shall see, in the modeling approach. Accordingly, if f_i is locally integrable, then $f_i(t, \mathbf{x}, v, \theta, u)\, v\, dv\, d\theta\, du\, d\mathbf{x}$ is the (expected) infinitesimal number of pedestrians whose microscopic state, at time t, is comprised in the elementary volume of the space of the microscopic states, corresponding to the variables space, velocity, and activity, of each functional subsystem.

In particular, the density and mean velocity of pedestrians belonging to the functional subsystem i reads

$$\rho_i(t, \mathbf{x}) = \int_0^1 \int_0^{2\pi} \int_0^1 f_i(t, \mathbf{x}, v, \theta, u)\, v dv\, d\theta\, du\,, \tag{6.9}$$

and

$$\boldsymbol{\xi}_i(t, \mathbf{x}) = \frac{1}{\rho_i(t, \mathbf{x})} \int_0^1 \int_0^{2\pi} \int_0^1 \mathbf{v}\, f_i(t, \mathbf{x}, v, \theta, u)\, v dv\, d\theta\, du\,, \tag{6.10}$$

whereas global expressions are obtained by summing over the index labeling the functional subsystems

$$\rho(t, \mathbf{x}) = \sum_{i=1}^{n} \rho_i(t, \mathbf{x}), \tag{6.11}$$

and

$$\boldsymbol{\xi}(t, \mathbf{x}) = \frac{1}{\rho(t, \mathbf{x})} \sum_{i=1}^{n} \rho_i(t, \mathbf{x}) \boldsymbol{\xi}_i(t, \mathbf{x}) \cdot \tag{6.12}$$

The kinetic theory approach suggests to refer the f_i to the maximal packing density n_M, so that ρ is a dimensionless density with values in $[0, 1]$. Further, dimensionless quantities are used as indicated in Remark 6.1.

An additional quantity to be taken into account in the modeling of interactions is the *perceived density* ρ_θ along the direction θ. According to [27], this quantity is defined as follows:

$$\rho_\theta[\rho] = \rho + \frac{\partial_\theta \rho}{\sqrt{1 + (\partial_\theta \rho)^2}} \big[(1 - \rho)\, H(\partial_\theta \rho) + \rho\, H(-\partial_\theta \rho)\big], \tag{6.13}$$

where ∂_θ denotes the derivative along the direction θ, while $H(\cdot)$ is the Heaviside function, $H(\cdot \geq 0) = 1$, and $H(\cdot < 0) = 0$. Therefore, positive gradients increase the perceived density up to the limit $\rho_\theta = 1$, while negative gradients decrease it down to the limit $\rho_\theta = 0$ in a way that

$$\partial_\theta \rho \to \infty \Rightarrow \rho_\theta \to 1\,, \quad \partial_\theta \rho = 0 \Rightarrow \rho_\theta = \rho\,, \quad \partial_\theta \rho \to -\infty \Rightarrow \rho_\theta \to 0,$$

where the limit value $\rho_\theta = 1$ is consistent with the use of dimensionless quantities mentioned in Remark 6.1, while the below limit $\rho_\theta = 0$ is consistent with the physics of the system which requires positivity of ρ.

6.3.2 A Mathematical Structure

This subsection shows how the mathematical tools derived in Chapter 3 can be further specialized toward the modeling of crowd dynamics. The structure presented in this subsection offers a quite general framework which, however, can be further generalized as suggested by possible applications or simplified to obtain simple models that might already depict phenomena of interest for applications. The presentation of the aforementioned framework is followed by a critical analysis of the existing literature with indication of what has been achieved and what can be done within possible research programs.

According to the mathematical approach proposed in Chapter 3, we consider the interactions of a *candidate particle* in $\mathbf{x}$ with microscopic state $\mathbf{v}_*, u_*$ which can acquire, in probability, the microscopic state of the *test particle* in $\mathbf{x}$ with microscopic state $\mathbf{v}, u$ due to the interaction with a *field particle* in $\mathbf{x}^*$ with microscopic state $\mathbf{v}^*, u^*$. Likewise, the test particle loses its state in the interaction with the field particle.

It is supposed that an active particle may interact only with those particles which are contained in its *interaction domain* Ω, i.e., $\mathbf{x}^* \in \Omega$. The latter is closely related to the particle visibility domain, namely a circular sector, with radius R, symmetric with respect to the velocity direction being defined by the visibility angles Θ and $-\Theta$, see Section 3.5 of Chapter 3. Interactions of test and candidate particles with field particles are modeled by the following quantities:

- *Interactionrate*: A candidate h-particle interacts with a field k-particle with a frequency $\eta_{hk}[\mathbf{f}](\mathbf{x}, \mathbf{x}^*, \mathbf{v}_*, \mathbf{v}^*, u_*, u^*; \alpha, \Sigma)$. Analogous notation is used for the interaction rate between a test i-particle and a k-particle, with $\mathbf{v}, u$ in place of $\mathbf{v}_*, u_*$ and i in place of h.
- *Transitionprobabilitydensity*: The probability that a candidate h-particle falls into the state of a test i-particle due to the interaction with a field k-particle is given by the probability density $\mathscr{A}_{hk}^i[\mathbf{f}](\mathbf{v}_* \to \mathbf{v}, u_* \to u|\mathbf{x}, \mathbf{x}^*, \mathbf{v}_*, \mathbf{v}^*, u_*, u^*; \alpha, \Sigma)$. When the system does not allow transitions across functional subsystems, the following notation is used $\mathscr{A}_{ik}^i =: \mathscr{A}_{ik}$, while when interactions do not depend on the other subsystems one has $\mathscr{A}_{ii}^i =: \mathscr{A}_i$.

Both η_{hk} and $\mathscr{A}_{hk}^i$ can depend on the microscopic states of the interacting particles and on the density of the field particles in the domain Ω.

The general mathematical structure which gives the time dynamics of the distribution functions f_i can be obtained, as shown in Chapter 3, by a balance of particles in the elementary volume of the space of the microscopic states:

Variation rate of the number of active particles
= Inlet flux rate - Outlet flux rate,

where the inlet and outlet fluxes are caused by interactions.

More specifically, this balance writes:

$$
\begin{aligned}
(\partial_t + \mathbf{v} \cdot \nabla_{\mathbf{x}})\; f_i(t, \mathbf{x}, \mathbf{v}, u) &= J_i[\mathbf{f}](t, \mathbf{x}, \mathbf{v}, u; \alpha, \Sigma) \\
&= \big(G_i - L_i\big)[\mathbf{f}](t, \mathbf{x}, \mathbf{v}, u; \alpha, \Sigma),
\end{aligned}
\tag{6.14}
$$

where J_i corresponds to Eq. (3.51) of Chapter 3. This structure is reported, when walkers do not move across functional subsystems, with tutorial aims as follows:

$$
\begin{aligned}
(\partial_t + \mathbf{v}\cdot\nabla_{\mathbf{x}})\, f_i(t,\mathbf{x},\mathbf{v},u) &= \sum_{k=1}^{n} \int_{(\Omega\times D_u^2\times D_{\mathbf{v}}^2)[\mathbf{f}]} \eta_{ik}[\mathbf{f}](\mathbf{x},\mathbf{x}^*,\mathbf{v}_*,\mathbf{v}^*,u_*,u^*)\\
&\quad\times \mathscr{A}_{ik}[\mathbf{f}](\mathbf{v}_*\to\mathbf{v},\, u_*\to u|\mathbf{x},\mathbf{x}^*,\mathbf{v}_*,\mathbf{v}^*,u_*,u^*)\\
&\quad\times f_i(t,\mathbf{x},\mathbf{v}_*,u_*) f_k(t,\mathbf{x}^*,\mathbf{v}^*,u^*)\, d\mathbf{x}^*\, d\mathbf{v}_*\, d\mathbf{v}^*\, du_*\, du^*\\
&\quad - f_i(t,\mathbf{x},\mathbf{v},u)\sum_{k=1}^{n}\int_{(\Omega\times D_u\times D_{\mathbf{v}})[\mathbf{f}]} \eta_{ik}[\mathbf{f}](\mathbf{x},\mathbf{x}^*,\mathbf{v},\mathbf{v}^*,u,u^*)\\
&\quad\times f_k(t,\mathbf{x}^*,\mathbf{v}^*,u^*)\, d\mathbf{x}^*\, d\mathbf{v}^*\, du^*.
\end{aligned}
\tag{6.15}
$$

Remark 6.4 *Specific models can be derived by simplifying the general mathematical structure. As an example, the model reported in the next section assumes that the rules by which a walker modify the velocity depend on the macroscopic quantities of the system as a whole. Furthermore, interactions are dealt with as the ones in the Lorentz gas where a test particle is supposed to collide with a fixed scattering background. Finally, the transition probability density $\mathscr{A}_i$ can be viewed as sum of two contributions, the former modeling interactions between walkers and the latter modeling non-local interactions with the walls.*

More in general, if the dynamics across functional subsystems is allowed, the structure is as follows:

$$
\begin{aligned}
(\partial_t + \mathbf{v}\cdot\nabla_{\mathbf{x}})\, f_i(t,\mathbf{x},\mathbf{v},u) &= \sum_{h,k=1}^{n} \int_{(\Omega\times D_u^2\times D_{\mathbf{v}}^2)[\mathbf{f}]} \eta_{hk}[\mathbf{f}](\mathbf{x},\mathbf{x}^*,\mathbf{v}_*,\mathbf{v}^*,u_*,u^*)\\
&\quad\times \mathscr{A}_{hk}^{i}[\mathbf{f}](\mathbf{v}_*\to\mathbf{v},\, u_*\to u|\mathbf{x},\mathbf{x}^*,\mathbf{v}_*,\mathbf{v}^*,u_*,u^*)\\
&\quad\times f_h(t,\mathbf{x},\mathbf{v}_*,u_*) f_k(t,\mathbf{x}^*,\mathbf{v}^*,u^*)\, d\mathbf{x}^*\, d\mathbf{v}_*\, d\mathbf{v}^*\, du_*\, du^*\\
&\quad - f_i(t,\mathbf{x},\mathbf{v},u)\sum_{k=1}^{n}\int_{(\Omega\times D_u\times D_{\mathbf{v}})[\mathbf{f}]} \eta_{ik}[\mathbf{f}](\mathbf{x},\mathbf{x}^*,\mathbf{v},\mathbf{v}^*,u,u^*)\\
&\quad\times f_k(t,\mathbf{x}^*,\mathbf{v}^*,u^*)\, d\mathbf{x}^*\, d\mathbf{v}^*\, du^*.
\end{aligned}
\tag{6.16}
$$

Remark 6.5 *The structures defined in Eqs. (6.15) and (6.16) allow us to define more precisely what has been done in the literature as well as what should be done. In more detail, models proposed in [43, 44] refer to (6.15), and an example will be presented in the forthcoming Section 6.4. On the other hand, structure (6.16) provides a conceptual framework still to be exploited with the aim of modeling complex social interactions as we shall see in the next Subsection 6.7.1.*

6.4 Mathematical Models

This section aims at showing how specific models can be derived within the general framework presented in Section 6.3. Various models have been proposed in the literature [29, 43, 44] which have been derived within the structure (6.15), while the structure (6.16) has not been yet extensively used to model the dynamics of walkers moving across functional subsystems. In addition, specific models refer to the simplification of the structure (6.15) mentioned in Remark 6.4 models, as the one presented in Subsection 6.4.1, refers to this specific simplified structure.

This section aims at showing how a specific model has been derived. This presentation is followed, in Section 6.5, by some simulations which enlighten the predictive ability of the model. However, the dynamics which includes crossing functional subsystems is an important research perspective which will be further discussed in the last section of this chapter. Indeed, social dynamics is a feature that can substantially modify the rules followed by walkers to organize their movement.

These topics are presented in the next subsections, where the first one deals with modeling topics, while Subsection 6.4.2 introduces some preliminary reasonings on the modeling stress conditions, which might be, for instance, induced by incidents which can generate, as we shall see, significant modifications on the overall dynamics.

6.4.1 Modeling Crowds in Bounded and Unbounded Domains

Let us now provide a few more details on the modeling approach based on mathematical tools derived by the kinetic theory of active particles. This approach was introduced in [29] for a crowd in unbounded domains where a finite number of velocity directions was allowed. This pioneer approach has been subsequently developed in [43] by using a continuous velocity distribution and accounting for interactions with walls and obstacles. Further technical refinements were introduced in [44] toward validation of the model based on its ability to reproduce empirical data.

A common feature of these models is that walkers organize their dynamics by a decision process by which each walker selects, induced by different stimuli and interactions, first the direction of the motion and subsequently adjust the velocity to the local flow conditions. The presence of walls is non-locally detected by walkers whose velocity is modified accordingly.

This decision process is consistent with the factorization of the transition probability density modeling interactions as follows

$$\mathscr{A}_{ik}[\mathbf{f}](\mathbf{v}_* \to \mathbf{v}; \alpha, \Sigma) = \delta(u - u_0)\, \mathscr{A}^{v}_{ik}[\mathbf{f}](v_* \to v; \alpha, \Sigma)\, \mathscr{A}^{\theta}_{ik}[\mathbf{f}](\theta_* \to \theta; \alpha, \Sigma), \tag{6.17}$$

for a crowd where the activity variable is uniformly distributed, with value $u = u_0$ and it is not modified by interactions, and where the velocity has been decomposed into speed v and direction of motion θ.

Let us consider a crowd which, according to the modeling approach proposed in the preceding section, can be subdivided into n functional subsystems. The state of the overall system is described by the one-particle distribution functions $f_i = f_i(t, \mathbf{x}, \mathbf{v})$ with $i = 1, \ldots, n$. In addition, the simplified structure introduced in Remark 6.4

is used referring specifically to Ref. [44]. In more detail, the model refers to the following structure:

$$
\begin{aligned}
&(\partial_t + \mathbf{v}\cdot\nabla_{\mathbf{x}})\, f_i(t,\mathbf{x},\mathbf{v}) = \\
&\quad \eta_{\mathrm{A}} \left(\int_{\mathscr{V}} \mathscr{A}[\rho,\boldsymbol{\xi}](\mathbf{v}_* \to \mathbf{v};\alpha) f_i(t,\mathbf{x},\mathbf{v}_*)\, d\mathbf{v}_* - f_i(t,\mathbf{x},\mathbf{v}) \right) \\
&\quad + \eta_{\mathrm{B}}\, d(\mathbf{x}) \left(\int_{\mathscr{V}} \mathscr{B}(\mathbf{v}_* \to \mathbf{v};\Sigma) f_i(t,\mathbf{x},\mathbf{v}_*)\, d\mathbf{v}_* - f_i(t,\mathbf{x},\mathbf{v}) \right), \qquad (6.18)
\end{aligned}
$$

where η_{A} and η_{B} are constant interaction rates between walkers and between walker and walls, $d(\mathbf{x})$ is a function defined in such a way that $d(\mathbf{x}) = 1$ within a domain at a distance d_w from the wall and $d(\mathbf{x}) = 0$ otherwise, and $\mathscr{A}$ and $\mathscr{B}$ are the transition probability densities which model the decision process based on which walkers modify their velocity. It is worth stressing that the dynamics depend only on $\mathbf{f}$ through the density ρ and the mean velocity ξ. Due to this feature, we can observe that the structure still presents important nonlinearities.

Let us now report the assumptions that have generated the model proposed in [44] accounting also for the rationale proposed in [43]. The modeling of each term of the mathematical structure is considered in detail:

Interaction rates: The interaction rate between walkers is assumed, for simplicity to be a constant, η_{A}. By contrast, it is supposed that walkers interact with walls only when they are sufficiently close to them. Accordingly, the interaction rate between walkers and walls is assumed to be space dependent, $\eta_{\mathrm{B}}\, d(\mathbf{x})$.

Velocity transition probability density: The dynamics by which walkers modify their velocity is given by two different terms, the former $\mathscr{A}$ modeling interaction between walkers and the latter $\mathscr{B}$ interaction with the walls.

Selection of the direction of motion: Three types of stimuli are supposed to contribute to the modification of walking direction, namely, the desire to reach a defined target, the attraction toward the mean stream, and the attempt to avoid overcrowded areas. These are represented by the three unit vectors $\boldsymbol{\nu}^{(t)}$, $\boldsymbol{\nu}^{(s)}$, and $\boldsymbol{\nu}^{(v)}$, respectively. It is expected that at high density, walkers try to drift apart from the more congested area moving in the direction of $\boldsymbol{\nu}^{(v)}$. Conversely, at low density, walkers head for the target identified by $\boldsymbol{\nu}^{(t)}$ unless their level of anxiety is high in which case they tend to follow the mean stream as given by $\boldsymbol{\nu}^{(s)}$. Accordingly, the preferred velocity direction is defined as

$$
\boldsymbol{\nu}^{(p)} = \frac{\rho\boldsymbol{\nu}^{(v)} + (1-\rho)\dfrac{\beta\boldsymbol{\nu}^{(s)} + (1-\beta)\boldsymbol{\nu}^{(t)}}{\left\|\beta\boldsymbol{\nu}^{(s)} + (1-\beta)\boldsymbol{\nu}^{(t)}\right\|}}{\left\|\rho\boldsymbol{\nu}^{(v)} + (1-\rho)\dfrac{\beta\boldsymbol{\nu}^{(s)} + (1-\beta)\boldsymbol{\nu}^{(t)}}{\left\|\beta\boldsymbol{\nu}^{(s)} + (1-\beta)\boldsymbol{\nu}^{(t)}\right\|}\right\|}, \qquad (6.19)
$$

where

$$\boldsymbol{\nu}^{(v)} = -\frac{\nabla_{\mathbf{x}}\rho}{\|\nabla_{\mathbf{x}}\rho\|}, \qquad \boldsymbol{\nu}^{(s)} = \frac{\boldsymbol{\xi}}{\|\boldsymbol{\xi}\|}. \tag{6.20}$$

In Eq. (6.19), $\beta \in [0, 1]$ is a parameter which models the sensitivity to the stream with respect to the search of vacuum and it is supposed modeling, to some extent, the level of anxiety of pedestrians. In general, this parameter might be related to the activity variable, namely $\beta = \beta(u)$. This issue is discussed later focusing on a preliminary analysis of panic conditions.

The transition probabilities for angles are defined as follows:

$$\mathscr{A}^{\theta}[\rho, \boldsymbol{\xi}](\theta_* \to \theta; \beta) = \delta\left(\theta - \theta^{(p)}[\rho, \boldsymbol{\xi}]\right), \tag{6.21}$$

where the preferred angle of motion, $\theta^{(p)}$, is obtained from Eq. (6.19) through the relation $\boldsymbol{\nu}^{(p)} = (\cos\theta^{(p)}, \sin\theta^{(p)})$.

Adjustment of the speed to local conditions: Two cases are distinguished:

- The walker's speed is greater than (or equal to) the mean speed. The walker either maintains its speed or decelerate to a speed ξ_d which is as much lower as density becomes higher. It is reasonable to assume that the probability to decelerate, p_d, increases with the congestion of the space, the quality of the venue and the anxiety level of the walker measured by the parameters α and β, respectively.
- The walker's speed is lower than the mean speed. The walker either maintains its speed or accelerate to a speed ξ_a which is as much higher as density becomes lower, the higher is the gap between the mean speed and the preferred speed and the goodness are the environmental conditions. It is reasonable to assume that the probability to accelerate, p_a, decreases with the congestion of the space and with the badness of the environmental condition and the anxiety level of the walker.

Accordingly, the transition probability density for the speed reads

$$\begin{aligned}\mathscr{A}^{v}[\rho, \boldsymbol{\xi}](v_* \to v; \alpha) &= \{p_a\delta\,(v - \xi_a) + (1 - p_a)\,\delta\,(v - v_*)\}\,H\,(\xi - v_*)\\ &\quad + \{p_d\delta\,(v - \xi_d) + (1 - p_d)\,\delta\,(v - v_*)\}\,H\,(v_* - \xi)\,,\end{aligned} \tag{6.22}$$

where

$$\xi_d = \xi - \rho_\theta^{(p)}\xi, \qquad p_d = (1 - c\,\alpha\,\beta)\rho_\theta^{(p)}, \tag{6.23}$$

and

$$\xi_a = \xi + \alpha\,\beta(1 - \rho_\theta^{(p)})(\alpha\,\beta - \xi), \qquad p_a = \alpha\,\beta(1 - \rho_\theta^{(p)}). \tag{6.24}$$

In Eqs. (6.23) and (6.24), $\rho_\theta^{(p)}$ is the perceived density, as given by Eq. (6.13), along the preferred walking direction $\theta^{(p)}$ and the constant $c < 1$ has been introduced so as to take into account that the probability to decelerate is different from zero even if $\alpha\,\beta = 1$.

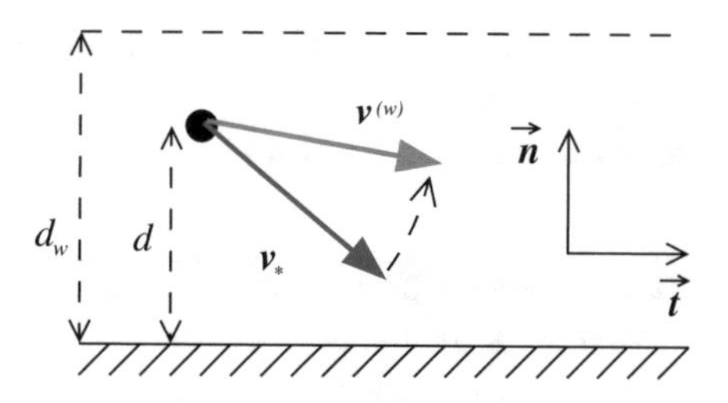

Fig. 6.1 Velocity adjustment in the presence of a wall

Modeling interactions with the walls: The modeling of the interactions between walkers and walls need to properly take into account its non-locality, as walkers are not classical particles and modify their velocity before encountering the wall. Accordingly, it is supposed that walkers whose distances from the wall, d, are within a specified cutoff, d_w, and whose velocity is directed to the wall, $\mathbf{v}_* \cdot \mathbf{n} < 0$, modify their velocity $\mathbf{v}_*$ to a new velocity $\mathbf{v}^{(w)}$, by reducing the normal component linearly with the distance from the wall but keeping the speed constant, that is

$$\mathbf{v}^{(w)} = \frac{d}{d_w}(\mathbf{v}_* \cdot \mathbf{n})\,\mathbf{n} + \mathrm{sign}(\mathbf{v}_* \cdot \mathbf{t}) \left[v_*^2 - \frac{d^2}{d_w^2}(\mathbf{v}_* \cdot \mathbf{n})^2 \right]^{1/2} \mathbf{t}, \tag{6.25}$$

where $\mathbf{n}$ and $\mathbf{t}$ are the normal and tangent to the wall. Figure 6.1 visualizes this specific dynamics.

The transition probability density $\mathscr{B}$ is then defined as

$$\mathscr{B}(\mathbf{v}_* \to \mathbf{v}; \Sigma) = \delta\left(\theta - \theta^{(w)}(d_w)\right) \delta\left(v - v^{(w)}\right), \tag{6.26}$$

where $v^{(w)} = v_*$ and $\theta^{(w)}$ is the direction of the velocity $\mathbf{v}^{(w)}$ defined by Eq. (6.25) through the relation $\mathbf{v}^{(w)}/v^{(w)} = (\cos\theta^{(w)}, \sin\theta^{(w)})$.

The overall modeling process is summarized in the following Tables 6.1 and 6.2.

6.4.2 Modeling Panic Conditions

When pedestrians move under stressful conditions, for instance during the evacuation from a congested area, they do not equally share the different trends but neglect the search of less congested areas and try to do what the others do [2]. Furthermore, the walking speed increases. This behavior reduces safety and can generate incidents.

In the models presented in the previous section, these aspects are properly accounted for by the parameter β. Indeed, in Eq. (6.19), this parameter weights the relative importance of the attraction toward the stream and, according to Eq. (6.22), the higher is β, the higher is the probability that pedestrians move with higher velocity.

Table 6.1 Pedestrian dynamics and related parameters

Pedestrian dynamics	*Parameter*
Interactions among pedestrians.	η_A, η_B : Interaction rates, i.e., number per unit of time and space of pedestrians that change their velocity.
Three stimuli modify the velocity direction: 1. desire to reach a well-defined target; 2. attraction toward the mean stream; 3. attempt to avoid overcrowded areas. The speed is modified depending on the **perceived density** as well as the quality of the environment and the level of walkers' stressful conditions.	α : Venue quality, which accounts for positive or negative slopes, good or bad lighting conditions, and so on. β : Stressful level, which weights relative importance of following the stream with respect to the search of vacuum.
Presence of walls adjusts the velocity direction by reducing its normal component linearly with the distance from the wall.	d_w : Cut-off distance, i.e., distance from the wall below which the presence of the wall has to be accounted for.

Table 6.2 Walker's decision process. Firstly walkers change the direction of movement and, afterwards, they modify their speed in probability. Walkers, who are close to a wall, reduce the velocity component normal to the wall linearly with the distance from the wall itself, keeping the speed constant

	Condition	*Transition*	*Probability*
Interactions	$\forall\,\theta_*$	$\theta_* \to \theta = \theta^{(p)}[\rho, \boldsymbol{\xi}]$	1
	$v_* \leq \xi$	$v_* \to v = \xi + \alpha\,\beta(1 - \rho_\theta^{(p)}[\rho])(\alpha\,\beta - \xi)$	$\alpha\,\beta(1 - \rho_\theta^{(p)}[\rho])$
		$v_* \to v = v_*$	$1 - \alpha\,\beta(1 - \rho_\theta^{(p)}[\rho])$
	$v_* > \xi$	$v_* \to v = \xi - \rho_\theta^{(p)}[\rho]\xi$	$(1 - c\alpha\,\beta)\rho_\theta^{(p)}[\rho]$
		$v_* \to v = v_*$	$1 - (1 - c\alpha\,\beta)\rho_\theta^{(p)}[\rho]$
Boundary	$d_* < d_w$	$\theta \to \theta = \theta^{(w)}(d_w)$ $v_* \to v = v_*$	1

If the crowd is in a sufficiently small domain, one can suppose that at a certain critical time t_c the aforesaid effects are homogeneously captured by the whole population. Therefore, the simplest approach is the following

$$t \leq t_c \; : \; \beta = \beta_1, \tag{6.27}$$

and

$$t > t_c \; : \; \beta = \beta_2, \tag{6.28}$$

where $0 < \beta_1 < \beta_2 < 1$.

On the other hand, in case of overcrowding in a large environment, one can figure out situations where panic, or any anomalous behavior, is initially localized in a small area and is then transported in the whole domain. The modeling approach can be developed by adding u to the set, which defines the microscopic state so that the distribution functions are $f_i = f_i(t, \mathbf{x}, v, \theta, u)$ with $\beta = \beta(u)$, while the transition probability density modifies as follows:

$$\mathscr{A}[\mathbf{f}](\mathbf{v}_* \to \mathbf{v}, u_* \to u; \alpha) = \mathscr{A}^u[\mathbf{f}](u_* \to u)\mathscr{A}^v[\rho, \boldsymbol{\xi}](v_* \to v; \alpha)\mathscr{A}^\theta[\rho, \boldsymbol{\xi}](\theta_* \to \theta; \alpha) \tag{6.29}$$

The modeling of $\mathscr{A}^u$ can be based on a consensus and learning type games, see Section 3 of [47] and, as an example, it may be related to the distance of the states of two interacting particles, see [46, 107].

It is plain that this very simple way to account for panic feelings in crowd dynamics does not exhaustively treat this topic whose importance is clear given the significant impact on safety of citizens [209–211]. As mentioned in Section 6.2, understanding the psychology of crowds in different circumstances needs to be transferred into mathematical structures. However, this effort still needs research activity to provide exhaustive results.

6.5 Crowd Dynamics Simulations

The effort to obtain numerical solutions of the model presented in Section 6.4 is definitely greater than that required by the kinetic equations presented in Chapter 5. In fact, the presence of the space variable and non-local interactions generates a challenging task because the distribution function depends on a high number of variables and the computation of the interaction terms requires the approximate evaluation of multidimensional integrals. An additional difficulty is induced by the fact that, in various cases such as applications related to safety problems, the requirement of real-time computation has to be fulfilled.

The numerical simulations discussed in the following subsections have been carried out by using a particle method which closely resemble the Direct Simulation Monte Carlo (DSMC) scheme first introduced by Bird [64]. The latter is by far the most popular and widely used simulation method in rarefied gas dynamics (see Refs. [117, 118] for recent applications to vacuum technology). This method has been subsequently further developed by various authors by modifying it to each specific system under consideration. Computational schemes for the specific application to crowd dynamics have been developed in [44]. Weighted particles can also be used to increase the results accuracy with only a small additional computational cost [207]. However, results obtained in the following have been obtained using the simplest approach which consists in representing each walker by a high number of computational particles, i.e., $10^5 - 10^6$. Numerical experiments have shown that this choice leads to percentage errors on estimated quantities which do not exceed 5% while the computing time is kept relatively low, being about 4–5 times higher than the effective simulated time.

This approach differs from deterministic methods of solution which adopt a completely different strategy, as the distribution function is discretized on a regular grid in the phase space. Then, the streaming term is approximated by finite volume schemes, and deterministic integration quadratures are used to evaluate the interaction integral [126]. However, compared to the deterministic approach, the particle method of solution permits to easily account for sophisticated individual decision processes and to handle venues with complex geometries while keeping the computing effort requirements at a reasonable level. Additional details on this computational approach have already been given in Chapter 2.

In the next two subsections, sample simulations are presented with a twofold aim; first, to prove that the model has the ability to reproduce the fundamental diagram, namely the dependence of the crowd speed from the density in steady and space homogeneous conditions; second, to show that the model can depict emerging behaviors which are observed in reality. Some results are of interest in important field of applications of crowd dynamics such as evacuation dynamics through venues with complex geometry.

6.5.1 Consistency with the Velocity Diagrams

The experimental study on crowd behavior is not developed as widely as the analogous field of vehicular traffic flows. Nevertheless, some empirical data have been reported in the literature, often in the form of speed and flow fundamental diagrams, see for instance Refs. [88, 216, 217].

The density–speed diagram provides the mean speed of walkers as a function of the local mean density in homogeneous steady conditions. In free flow conditions, walkers move at the average maximum allowed speed which depends on the environmental conditions. Conversely, in congested flow conditions, walkers move closer to one another at a reduced speed, until the density reaches the jam density at which walkers stop and have zero speed.

It is worth stressing that the aforementioned diagram should not be artificially inserted into the equations of the model but rather be reproduced as a consequence of interactions at the microscopic scale. This is the modeling approach developed in Ref. [111], where, starting from a statistical description of the microscopic interactions between vehicles, the fundamental diagrams of traffic flow have been obtained as stationary long-time asymptotic solutions of the corresponding kinetic equations. Likewise, we here show that, in the space-homogeneous case, the proposed model admits a closed form solution of the density–speed behavior.

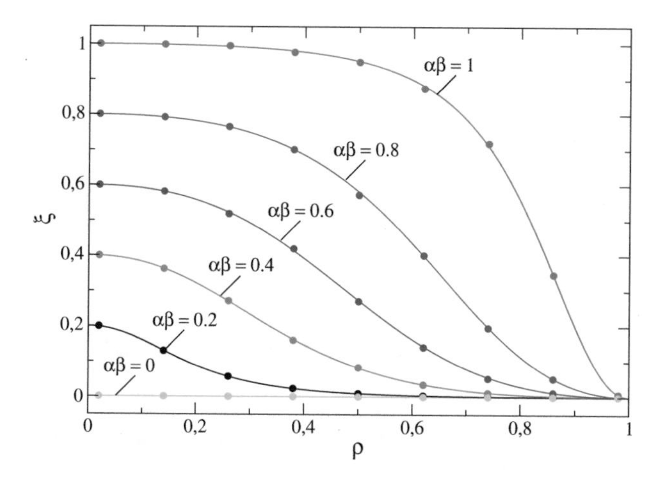

Fig. 6.2 Fundamental diagram as a function of $\alpha\beta$. Solid lines are the analytical solution as given by Eq. (6.30) with $c = 0.95$. Solid symbols are the numerical predictions obtained by the Monte Carlo method

Indeed, by referring to a flow which is one-dimensional in space and time independent, a closed form solution of the kinetic model can be sought in the form $f(t, \mathbf{x}, \mathbf{v}) = f^{(eq)}(v)$, of the model. Without repeating simple technical calculations developed in [44], let us report the final result that provides the mean speed versus density depending on the parameters of the model:

$$\xi = \frac{\alpha^3\,\beta^3\,(1-\rho)^2}{\alpha^2\,\beta^2\,(1-\rho)^2 + (1 - c\,\alpha\,\beta)\,\rho^2}\,. \tag{6.30}$$

It is worth noticing that Eq. (6.30) depends on the quality of the venue, α, and the trend of walkers to imitate their surroundings instead of searching less crowded areas, β, only through their product $\gamma = \alpha\beta$.

The fundamental diagram as provided by Eq. (6.30) is shown in Fig. 6.2. The mean features of the empirical evidence about the relation between density and speed in a crowd are correctly reproduced. More specifically, the mean velocity of walkers decreases monotonically with the density from the maximal value for $\rho = 0$ to zero when $\rho \to 1$. Moreover, the maximal velocity observed at very low density increases with the parameter γ, namely with the quality of the environmental conditions and/or the walkers' anxiety. The slope of the decay also depends on γ. In more detail, anxious walkers within high-quality venues show a trend to keep the maximal velocity up to high densities while the mean velocity reduces more rapidly to zero in the opposite case.

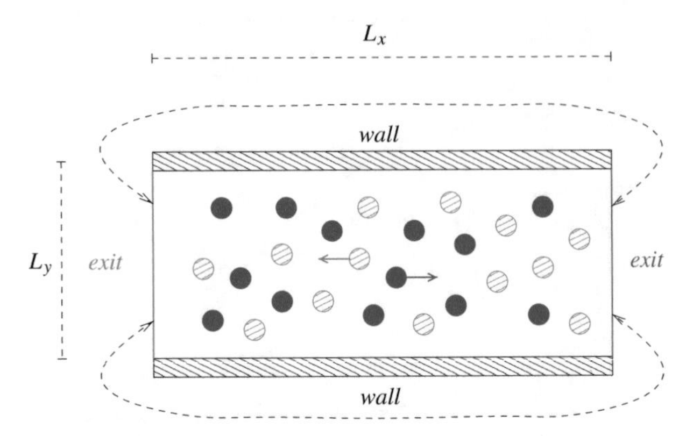

Fig. 6.3 Case study: Pedestrian counterflow in a narrow street. Periodic boundary conditions are assumed in the longitudinal direction

6.5.2 *Self-organized Lane Formation*

Let us consider the two-dimensional space non-homogeneous dynamics of self-organized lane formation, where two groups of pedestrians move toward opposite directions in a narrow street, see Fig. 6.3. The groups might tend to organize their movement into a number of dynamically varying lanes. Actually this behavior is shown by empirical data.

Simulations are carried out for a crowd composed by 50 pedestrians who are initially uniformly distributed over the domain. Beside describing the segregation of pedestrians into lanes of uniform walking direction, the influence of the parameter β on the overall dynamics is assessed.

The spontaneous formation of parallel lanes can be quantitatively evaluated by computing the band index, $Y_{\mathrm{B}}(t)$, which measures the segregation of opposite flow directions [43, 248] and reads

$$Y_{\mathrm{B}}(t) = \frac{1}{L_x L_y} \int_0^{L_y} \left| \int_0^{L_x} \frac{\rho_1(t,\mathbf{x}) - \rho_2(t,\mathbf{x})}{\rho_1(t,\mathbf{x}) + \rho_2(t,\mathbf{x})} \, dx \right| dy, \tag{6.31}$$

where $\mathbf{x} = \{x, y\}$.

According to its definition, one has $Y_{\mathrm{B}}(t) = 0$ for mixed counterflows and $Y_{\mathrm{B}}(t) = 1$ for a perfect segregation of the opposite flows. As shown in Fig. 6.4, the long-time value of the band index is a non-monotonic function of β, and the emergence of the spatial segregation is anticipated to be more pronounced for $\beta = 0.5$. By contrast, for higher values of β, the self-organization pattern is expected to disappear with the pedestrians randomly filling the domain.

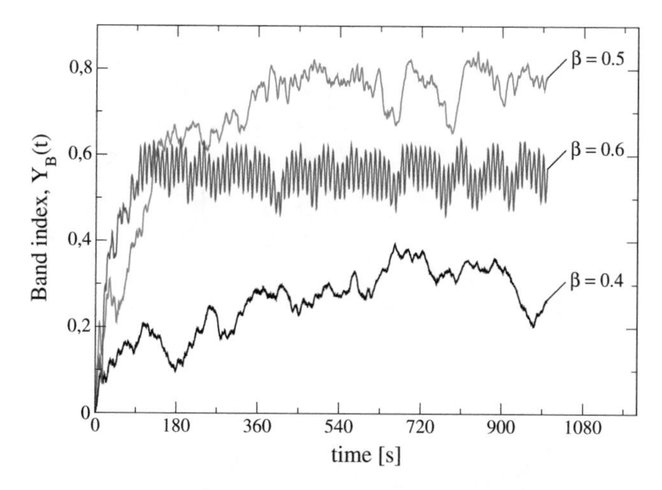

Fig. 6.4 Band index of the density field related to the counterflow of a crowd composed by 50 pedestrians for different β

These results are clearly shown in Figs. 6.5 and 6.6. It is worth pointing out that, unlike most of the previous studies on the subject, this emerging behavior has not been obtained by introducing a repulsion force between pedestrians belonging to different populations.

6.5.2.1 Evacuation Dynamics from a Complex Venue

Let us now consider the evacuation from a complex venue. It is expected that the time elapsed between the instant that pedestrians receive the warning to evacuate and their arrival at destination increases under stressful condition.

A sketch of the initial conditions and of the geometry of the venue are shown in Fig. 6.7. The venue consists in three rooms connected by the doors d_1, d_2, and d_3, while the exit is located in the upper right corner at door d_4. In detail, we consider the following case study.

A group of walkers is concentrated in a circular area, of radius $4\,m$, initially just standing. Then they receive an evacuation information; the original group symmetrically divides into two groups moving toward the closest between the doors d_1 and d_2. Walkers who do not get through door d_1 before it suddenly closes at $t = 20s$ are obliged to change direction and move to door d_2 in the attempt to reach door d_4. As in the preceding subsection, special focus is assigned to the role of the parameter β which is descriptive of the presence of stressful conditions.

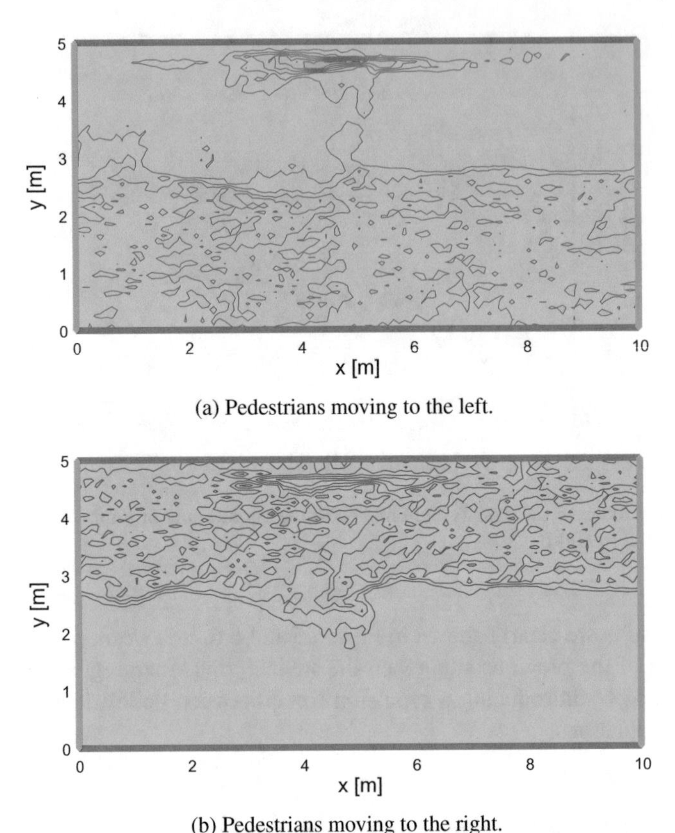

(a) Pedestrians moving to the left.

(b) Pedestrians moving to the right.

Fig. 6.5 Density contour plot of a crowd composed by 50 pedestrians divided into two groups which move leftward and rightward for $\beta = 0.5$

Let us stress that these simulations can contribute to crisis managing not only by showing that the model can provide a realistic description of the crowd dynamics, but also by verifying how the evacuation time increases under stressful conditions and the identification of risk situation due to an excessive concentration of walkers in the same area. Indeed, safety conditions require that local density remains below a safety level. This dynamics depends on venues, hence simulations, as mentioned, can contribute to optimize the shape of the areas, where walkers move.

Figures 6.8 and 6.9 show the patterns of the crowd flow at different times before and after the closure of the door caused by an hypothetical incident. The closure causes the onset of stress modeled by an increase of the parameter β. This event generates a dangerous overcrowding on the doors of the upper part of the venues and a loss of orientation of some walkers.

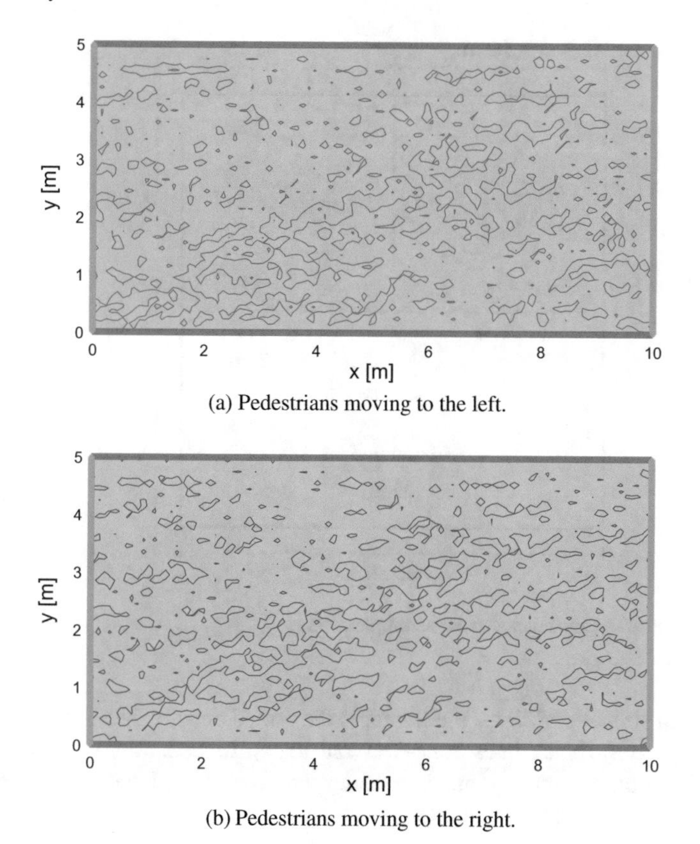

(a) Pedestrians moving to the left.

(b) Pedestrians moving to the right.

Fig. 6.6 Density contour plot of a crowd composed by 50 pedestrians divided into two groups which move leftward and rightward for $\beta = 0.8$

Figure 6.10 shows the ratio between the walkers in the room and their initial number versus time for different density of the crowd and with/without the increase of stress after the incident occurrence.

These simulations only cover a small part of the overall study which, thank to several possible conditions and parameter sensitivity analysis, has allowed to draw the following conclusions:

- Independently on the crowd density, at the first stage of the evacuation process, the higher is β, the faster is the evacuation time (dashed lines are above solid lines). This is not unexpected since the walkers mean velocity increases with β. However in the long run, stress conditions increase the evacuation time (dashed lines are below solid lines).

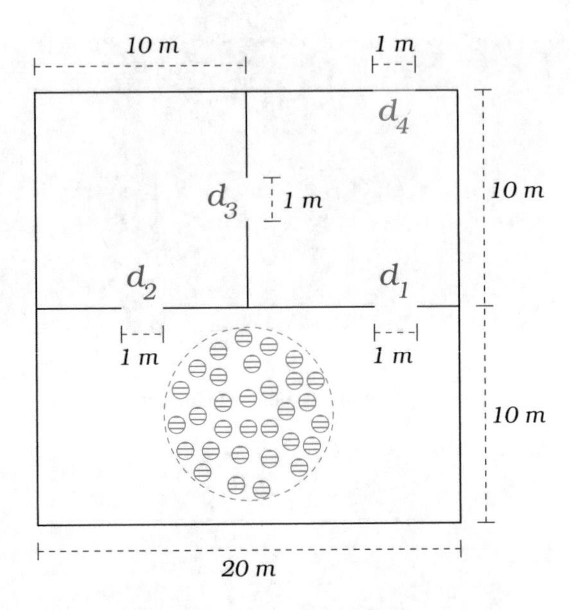

Fig. 6.7 Case study: Evacuation from a complex venue

- For equal stress conditions, the higher is the crowd density the slower is the evacuation process (black lines are below red lines). Indeed, area with high density arise from the dynamics, thus increasing the congestion level and consequently reducing the local mean velocity.

6.6 On the Dynamics of Multicellular Systems

This section presents an application of the KTAP approach to modeling and simulations of multicellular systems. The general idea, which is here developed, is quite different from that presented in the preceding section as it consists in modeling the dynamics as a perturbation of the spatially homogeneous behavior, see Subsection 3.5.6. This approach, although it is less general than that used for crowd modeling, presents the advantage of leading to the derivation of macroscopic type models which can be solved by classical deterministic methods of numerical analysis.

In more detail, we refer to [195], where this approach has been applied to the modeling of multicellular systems in which the asymptotic analysis refers to the hyperbolic limit for a dynamics somewhat related to chemotaxis and cross diffusion phenomena [33] which are generally modeled by parabolic equations. However, the hyperbolic scaling appears to be more appropriate to model biological phenomena where propagation occurs with finite speed.

The presentation is limited to a rapid description of modeling issues and of simulations followed by their interpretation. The reader interested to a more detailed

Fig. 6.8 Density contour plots of a crowd composed by 50 walkers evacuating the room at different instants of time before the closure of the door with $\beta = 0.5$

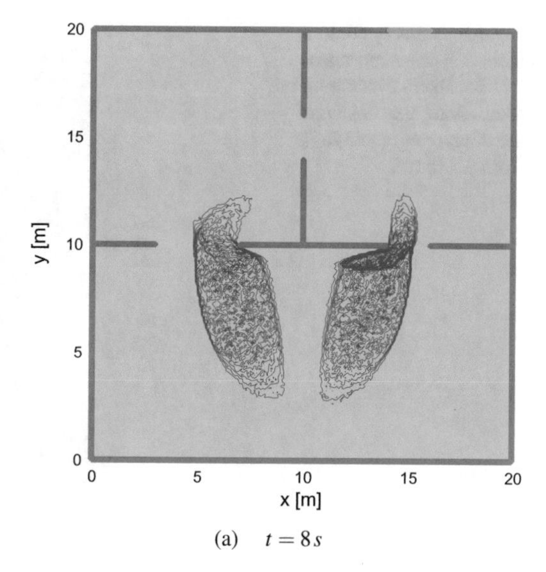

(a) $t = 8\,s$

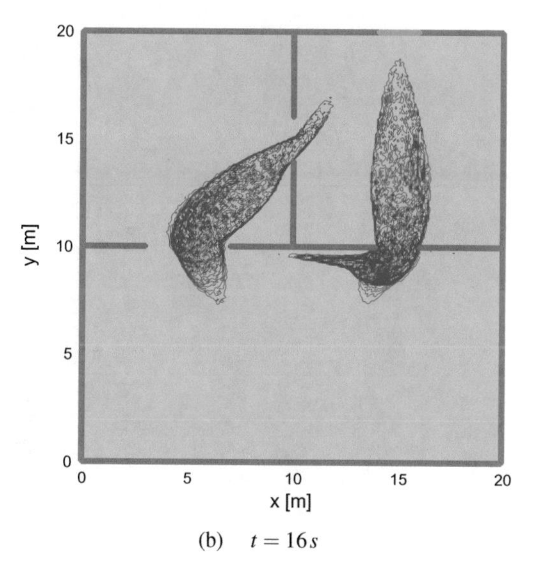

(b) $t = 16\,s$

description is referred to [195] and to [30] for the derivation of macroscopic equations from the underlying description at the microscopic scale delivered by kinetic theory models, while the mathematical theory of Keller–Segel models is available in the survey [33]. General aspects of micro–macro derivation are treated in Chapter 7, where a broad literature is brought to the attention of the interested reader.

The presentation is in two steps treated in the next subsections. Namely, the first subsection presents the kinetic model from which the macroscopic model is derived, while some simulations are presented in the second subsection.

Fig. 6.9 Density contour plots of a crowd composed by 50 walkers evacuating the room. After the closure of the door, panic modifies β from 0.5 to 0.8

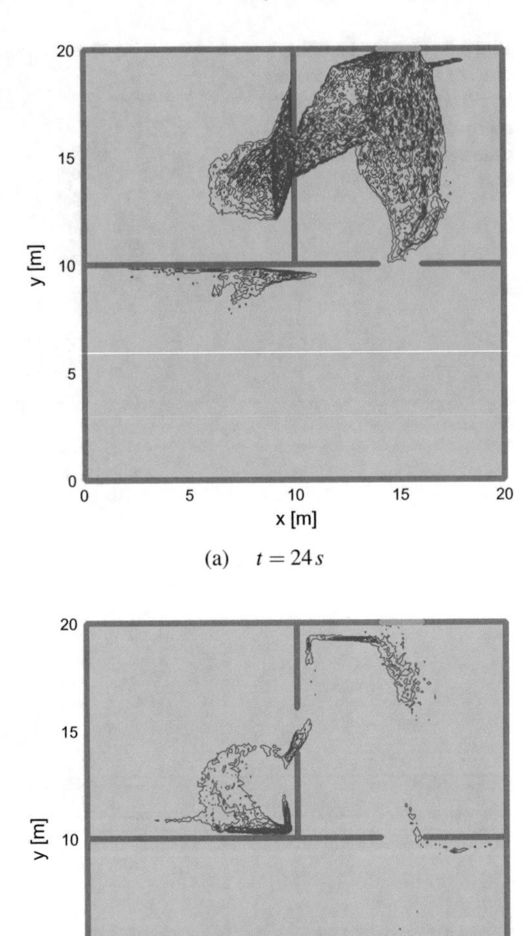

(a) $t = 24\,s$

(b) $t = 40\,s$

6.6.1 *From Kinetic Theory to a Macroscopic Model*

Let us now consider a physical system constituted by a large number of cells interacting in a biological environment. The microscopic state is defined by the mechanical variable

$$\{\mathbf{x}, \mathbf{v}\} \in \Omega \times V \subset \mathbb{R}^d \times \mathbb{R}^d, \qquad d = 1, 2, 3.$$

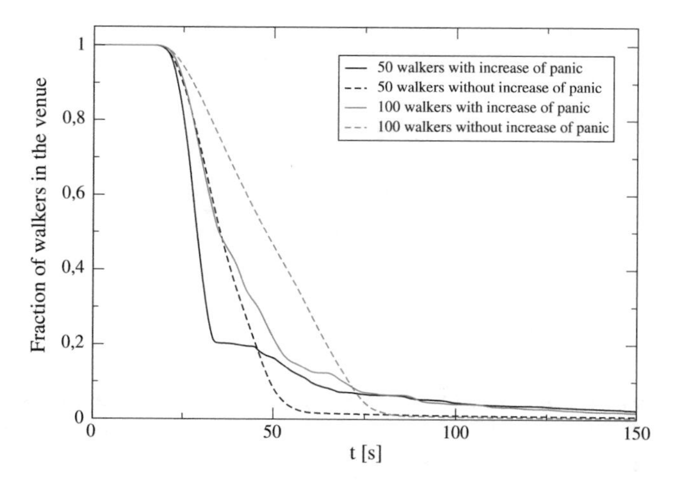

Fig. 6.10 Fraction of walkers in the room versus time. Evacuation of a crowd composed of 50 (black lines) and 100 (red lines) walkers, when the parameter β modifies from 0.5 to 0.8 (solid lines) and remains at the constant value of 0.5 (dashed lines)

The statistical collective description of the system is encoded in the distribution function

$$f = f(t, \mathbf{x}, \mathbf{v}) : [0, T] \times \Omega \times V \to \mathbb{R}_+.$$

The system treated in [195] consists of different species in response to multiple chemotactic cues in which the density (concentration) of each chemotactic cue is denoted by $g_i = g_i(t, \mathbf{x}, \mathbf{v}) : [0, T] \times \Omega \times V \to \mathbb{R}_+$ for $i = 1, \dots, m$ and m is the total number of the chemotactic cues. It is also assumed that the transport in position is linear with respect to the velocity. Then, the evolution of the distribution functions f and g_i can be modeled as follows:

$$\begin{cases} \partial_t f + \mathbf{v} \cdot \nabla_{\mathbf{x}} f = L(\mathbf{g}, f) + \widetilde{H}(f, \mathbf{g}), \\ \tau_i \, \partial_t g_i + \mathbf{v} \cdot \nabla_{\mathbf{x}} g_i = l_i(g_i) + G_i(f, \mathbf{g}), \end{cases} \tag{6.32}$$

where $\mathbf{g} = (g_1, \dots, g_m)^T$ and $\tau_i \in \mathbb{R}_+$ is a dimensionless time factor which indicates that the spatial spread of f and g_i is on different time scales. The operators L and l_i model the dynamics of biological organisms by velocity-jump process while $\widetilde{H}$ and G_i describe proliferation/destruction interactions.

The above model describes the evolution of a large system of interacting cells in response to multiple chemotactic cues. Interactions occur within the microscopic domain $\Omega \times V$ of the test cell. V is assumed to be bounded and radially symmetric.

The information at the macroscopic scale can be obtained by moment calculations, more precisely the zero-order moments are computed as follows:

$$n(t, \mathbf{x}) = \int_V f(t, \mathbf{x}, \mathbf{v})\, d\mathbf{v} \quad \text{and} \quad N_i(t, \mathbf{x}) = \int_V g_i(t, \mathbf{x}, \mathbf{v})\, d\mathbf{v}, \tag{6.33}$$

while the first-order moments are given by:

$$\xi(t, \mathbf{x}) = \frac{1}{n(t, \mathbf{x})} \int_V \mathbf{v}\, f(t, \mathbf{x}, \mathbf{v})\, d\mathbf{v}, \tag{6.34}$$

and

$$U_i(t, \mathbf{x}) = \frac{1}{N_i(t, \mathbf{x})} \int_V \mathbf{v} g_i(t, \mathbf{x}, \mathbf{v})\, d\mathbf{v}. \tag{6.35}$$

Equations for the moments (6.33), (6.34), and (6.35) can be obtained as a hydrodynamic limit after an appropriate scaling of time and space, different time-space scalings lead to equations characterized by different parabolic or hyperbolic structures, different combinations of parabolic and hyperbolic scales also are used, according to the dispersive or non-dispersive nature of the biological system under consideration. However, a more recent tendency has been the use of hyperbolic equations to describe intermediate regimes at the macroscopic level rather than parabolic equations, for example [30, 195]. In the next paragraph, it is shown how the hyperbolic models may be derived as a hydrodynamic limit of the kinetic equation (6.32).

Let us now consider the derivation of models at the macroscopic scale according to a hyperbolic scaling: $t \to \varepsilon t$ and $x \to \varepsilon x$, applied to the first equation of system (6.32), where ε is a small parameter which will be allowed to tend to zero [30]. In addition, a smallness assumption is made on interactions $\widetilde{H}(f, \mathbf{g}) = \varepsilon H(f, \mathbf{g})$.

The derivation is based on the assumption that L admits the following decomposition:

$$L(\mathbf{g}, f) = L^0(f) + \varepsilon L^1(\mathbf{g}, f), \tag{6.36}$$

with L^1 in the form

$$L^1(\mathbf{g}, f) = \sum_{i=1}^{m} L_i^1[g_i](f). \tag{6.37}$$

The operator L^0 represents the dominant part of the turning kernel modeling the tumble process in the absence of chemical substance, and L_i^1 is the perturbation due to chemical cues. With this considerations, Eq. (6.32) becomes:

$$\begin{cases} \partial_t f + \mathbf{v} \cdot \nabla_{\mathbf{x}} f = \dfrac{1}{\varepsilon} L^0(f) + \displaystyle\sum_{i=1}^{m} L_i^1[g_i](f) + H(f, \mathbf{g}), \\ \tau_i \partial_t g_i + \mathbf{v} \cdot \nabla_{\mathbf{x}} g_i = l_i(g_i) + G_i(f, \mathbf{g}). \end{cases} \tag{6.38}$$

Let us now consider the asymptotic limit of (6.38) as ε goes to zero. Some assumptions on the turning operators L^0, L_i^1, and l_i are necessary to develop this asymptotic analysis:

Assumption H0. (Conservation of the local mass) The turning operators L^0, L_i^1, and l_i conserve the local mass:

$$\int_V L^0(f)\,d\mathbf{v} = \int_V L_i^1[g_i](f)\,d\mathbf{v} = \int_V l_i(f)\,d\mathbf{v} = 0.$$

Assumption H1. (Conservation of the population flux) The turning operator L^0 conserves the population flux:

$$\int_V \mathbf{v}\,L^0(f)\,d\mathbf{v} = 0.$$

Assumption H2. (Kernel of L^0) For all $n \in [0, +\infty[$ and $\xi \in \mathbb{R}^d$, there exists a unique function $F_{n,\xi} \in L^1(V, (1 + |\mathbf{v}|)d\mathbf{v})$ such that

$$L^0(F_{n,\xi}) = 0, \quad \int_V F_{n,\xi}\,d\mathbf{v} = n \quad \text{and} \quad \int_V \mathbf{v}\,F_{n,\xi}\,d\mathbf{v} = n\,\xi.$$

Multiplying each equation of system (6.38) by 1 and $\mathbf{v}$ and integrating over V yields:

$$\begin{cases} \partial_t n + div_{\mathbf{x}}(n\,\xi) = \int_V H(f, \mathbf{g})\,d\mathbf{v}, \\ \partial_t(n\xi) + div_{\mathbf{x}} \int_V \mathbf{v} \otimes \mathbf{v} f(t, \mathbf{x}, \mathbf{v})\,d\mathbf{v} = \sum_{i=1}^{m} \int_V \mathbf{v} L_i^1[g_i](f)\,d\mathbf{v} + \int_V \mathbf{v}\,H(f, \mathbf{g})\,d\mathbf{v}, \\ \tau_i \partial_t N_i + div_x(N_i U_i) = \int_V G_i(f, \mathbf{g})\,d\mathbf{v}, \\ \tau_i \partial_t(N_i U_i) + div_{\mathbf{x}} \int_V \mathbf{v} \otimes \mathbf{v} g_i(t, \mathbf{x}, \mathbf{v})\,d\mathbf{v} = \int_V \mathbf{v}\,l_i(g_i)\,d\mathbf{v} + \int_V \mathbf{v} G_i(f, \mathbf{g})\,d\mathbf{v}. \end{cases} \tag{6.39}$$

Let us consider a small perturbation of the solution $f(t, \mathbf{x}, \mathbf{v})$ near the equilibrium to close system (6.39)

$$f(t, \mathbf{x}, \mathbf{v}) = F_{n(t,\mathbf{x}),\xi(t,\mathbf{x})}(\mathbf{v}) + \varepsilon f_1(t, \mathbf{x}, \mathbf{v}), \tag{6.40}$$

where the equilibrium distribution $F_{n,\xi}$ is given by Assumption H2. In addition, let us consider the following asymptotic expansion in order 1 in ε:

$$H(\phi + \varepsilon\psi, \theta) = H(\phi, \theta) + O(\varepsilon) \text{ and } G_i(\phi + \varepsilon\psi, \theta) = G_i(\phi, \theta) + O(\varepsilon). \tag{6.41}$$

Replacing now f by its expansion (6.40) in (6.39) yields:

$$\begin{cases} \partial_t n + div_{\mathbf{x}}(n\,\xi) = \int_V H(F_{n,\xi}, \mathbf{g})\, d\mathbf{v} + O(\varepsilon), \\ \partial_t(n\xi) + div_{\mathbf{x}}(P + n\xi \otimes \xi) = \sum_{i=1}^{m} \int_V (\mathbf{v} - \xi) L_i^1[g_i](F_{n,\xi}) d\mathbf{v} \\ \qquad + \int_V \mathbf{v} H(F_{n,\xi}, \mathbf{g})\, d\mathbf{v} + O(\varepsilon), \\ \tau_i \partial_t N_i + div_{\mathbf{x}}(N_i U_i) = \int_V G_i(F_{n,\xi}, \mathbf{g}) d\mathbf{v} + O(\varepsilon), \\ \tau_i \partial_t(N_i U_i) + div_{\mathbf{x}} \int_V \mathbf{v} \otimes \mathbf{v} g_i(t, \mathbf{x}, \mathbf{v})\, d\mathbf{v} = \int_V \mathbf{v} l_i(g_i) d\mathbf{v} \\ \qquad + \int_V \mathbf{v} G_i(F_{n,\xi}, \mathbf{g})\, d\mathbf{v} + O(\varepsilon), \end{cases} \tag{6.42}$$

where the pressure tensor P is given by

$$P(t, \mathbf{x}) = \int_V (\mathbf{v} - \xi(t, \mathbf{x})) \otimes (\mathbf{v} - \xi(t, \mathbf{x})) F_{n(t,\mathbf{x}),\xi(t,\mathbf{x})}(\mathbf{v})\, d\mathbf{v}. \tag{6.43}$$

The closure can be obtained by looking for an approximate expression of the second-order moment

$$\int_V \mathbf{v} \otimes \mathbf{v} g_i(t, \mathbf{x}, \mathbf{v})\, d\mathbf{v}.$$

The approach consists in deriving a function $a_i(t, \mathbf{x}, \mathbf{v})$ which minimizes the $L^2(V)$-norm under the constraints that it has the same first moments, N_i and $N_i U_i$, as g_i. Once a_i this function has been found, we replace g_i by a_i in Eq. (6.42).

This minimization problem can be solved explicitly in the case where the set of velocity is the sphere of radius $s > 0$, $V = s\mathbb{S}^{d-1}$, using the Lagrangian multipliers and one finds:

$$a_i(t, \mathbf{x}, \mathbf{v}) = \frac{1}{|V|} \left(N_i(t, \mathbf{x}) + \frac{d}{s^2} N_i(t, \mathbf{x}) U_i(t, \mathbf{x}) \cdot \mathbf{v} \right). \tag{6.44}$$

Thus, the following nonlinear coupled hyperbolic model, at first order with respect to ε, is derived

$$\begin{cases} \partial_t n + div_{\mathbf{x}}(n\xi) = \int_V H(F_{n,\xi}, a)\, d\mathbf{v} + O(\varepsilon), \\ \partial_t(n\xi) + div_{\mathbf{x}}(P + n\xi \otimes \xi) = \sum_{i=1}^{m} \int_V (v - \xi) L_i^1[a_i](F_{n,\xi})\, d\mathbf{v} \\ \qquad + \int_V v H(F_{n,\xi}, a) d\mathbf{v} + O(\varepsilon), \\ \tau_i \partial_t N_i + div_{\mathbf{x}}(N_i U_i) = \int_V G_i(F_{n,\xi}, a)\, d\mathbf{v} + O(\varepsilon), \\ \tau_i \partial_t(N_i U_i) + \frac{s^2}{d} \nabla_{\mathbf{x}} N_i = \int_V v l_i(a_i)\, d\mathbf{v} + \int_V \mathbf{v} G_i(F_{n,\xi}, a)\, d\mathbf{v} + O(\varepsilon), \end{cases} \tag{6.45}$$

with $a = (a_1, \cdots, a_m)$.

Specific models can be obtained by an appropriate specialization of the various terms in (6.45). In more detail, we consider the equilibrium function $F_{n,\xi}$ as follows:

$$F_{n,\xi}(\mathbf{v}) = \frac{1}{|V|}(n + \frac{d}{s^2} nu \cdot \mathbf{v}). \tag{6.46}$$

The perturbation operators L_0, L_i^1, and l_i are supposed to be integral operators as follows:

$$L^0(f) = \int_V \big(T^0(\mathbf{v}, \mathbf{v}') f(t, \mathbf{x}, \mathbf{v}') - T^0(\mathbf{v}', \mathbf{v}) f(t, \mathbf{x}, \mathbf{v})\big)\, d\mathbf{v}', \tag{6.47}$$

$$L_i^1[g_i](f) = \int_V \big(T_i^1(g_i, \mathbf{v}, \mathbf{v}') f(t, \mathbf{x}, \mathbf{v}') - T_i^1(g_i, \mathbf{v}', \mathbf{v}) f(t, \mathbf{x}, \mathbf{v})\big)\, d\mathbf{v}', \tag{6.48}$$

and

$$l_i(f) = \int_V \big(K_i(\mathbf{v}, \mathbf{v}') f(t, \mathbf{x}, \mathbf{v}') - K_i(\mathbf{v}', \mathbf{v}) f(t, \mathbf{x}, \mathbf{v})\big)\, d\mathbf{v}'. \tag{6.49}$$

The turning kernels $T^0(\mathbf{v}, \mathbf{v}')$, $T_i^1(g_i, \mathbf{v}, \mathbf{v}')$, and $K_i(\mathbf{v}, \mathbf{v}')$ describe the reorientation of cells, i.e., the random velocity changes from the previous velocity $\mathbf{v}'$ to the new $\mathbf{v}$, and are defined as follows:

$$T^0(\mathbf{v}, \mathbf{v}') = \frac{\mu_0}{|V|}(1 + \frac{d}{s^2} \mathbf{v} \cdot \mathbf{v}'),$$

$$T_i^1[g_i](\mathbf{v}, \mathbf{v}') = \frac{\mu_1}{|V|} - \frac{\mu_2 d}{|V| s^2} \mathbf{v}' \cdot \big(\alpha_i(< g_i >) \nabla_{\mathbf{x}}(< g_i >)\big),$$

and

$$K_i(\mathbf{v}, \mathbf{v}') = \frac{\sigma_i}{|V|},$$

where μ_0, μ_1, μ_2, and σ_i are real constants, α_i is a mapping $\mathbb{R} \longrightarrow \mathbb{R}$ and $< \cdot >$ stands for the $(\mathbf{v})$-mean of a function, i.e., $< h >:= \displaystyle\int_V h(t, \mathbf{x}, \mathbf{v})\, d\mathbf{v}$ for $h \in L^2(V)$. Thus, the hyperbolic system (6.45), detailed calculations can be found in [195], becomes

$$\begin{cases} \partial_t n + div_{\mathbf{x}}(n\xi) = \displaystyle\int_V H(F_{n,\xi}, a)\, d\mathbf{v}, \\ \partial_t(n\xi) + \dfrac{s^2}{d} \nabla_{\mathbf{x}} n = -\mu_1 m\, n\xi + \mu_2 \displaystyle\sum_{i=1}^{m} n\alpha_i(N_i) \nabla_{\mathbf{x}} N_i + \int_V \mathbf{v} H(F_{n,\xi}, a)\, d\mathbf{v}, \\ \tau_i \partial_t N_i + div_{\mathbf{x}}(N_i U_i) = \displaystyle\int_V G_i(F_{n,\xi}, a)\, d\mathbf{v}, \\ \tau_i \partial_t(N_i U_i) + \dfrac{s^2}{d} \nabla_{\mathbf{x}} N_i = -\sigma_i N_i U_i + \displaystyle\int_V \mathbf{v} G_i(F_{n,\xi}, a)\, d\mathbf{v}. \end{cases} \tag{6.50}$$

The derivation of a chemotaxis model can be deduced from the hyperbolic structure (6.50) in the case when $m = 1$. Indeed, let us assume the following scaling:

$$\sigma_1 \to \infty, \quad s \to \infty, \quad \mu_1 = \mu_2, \quad \text{such that} \quad \frac{s^2}{d\sigma_1} \to D_{N_1} \quad \text{and} \quad \frac{s^2}{d\mu_1} \to D_n,$$

and suppose that

$$H(F_{n,u}, a_1) = H\left(\frac{n}{|V|}, \frac{N_1}{|V|}\right) + O(\frac{1}{s^2}) \text{ and } G_1(F_{n,u}, a_1) = G_1\left(\frac{n}{|V|}, \frac{N_1}{|V|}\right) + O(\frac{1}{s^2}).$$

Now, dividing the second and fourth equations in (6.50) by μ_1 and σ_1, respectively, and taking last limits, the following model is obtained:

$$\begin{cases} \partial_t n = div_{\mathbf{x}}(D_n \nabla_{\mathbf{x}} n - n\alpha_1(N_1)\nabla_{\mathbf{x}} N_1) + \widetilde{H}(n, N_1), \\ \tau_1 \partial_t N_1 = D_{N_1} \Delta_{\mathbf{x}} N_1 + \widetilde{G_1}(n, N_1), \end{cases} \tag{6.51}$$

where,

$$\widetilde{H}(n, N_1) = |V| H\left(\frac{n}{|V|}, \frac{N_1}{|V|}\right) \quad \text{and} \quad \widetilde{G_1}(n, N_1) = |V| G_1\left(\frac{n}{|V|}, \frac{N_1}{|V|}\right).$$

System (6.51) consists of two coupled reaction–diffusion equations, which are parabolic equations. Moreover, this model is one of the simplest models to describe the aggregation of cells by chemotaxis [33].

6.6.2 A Numerical Test and Critical Analysis

This subsection presents some numerical simulations, with the choice $m = 1$, $H = 0$, $G_1 = \frac{n}{|V|}$, and $\tau_1 = 1$, in a two-dimensional setting, to enlighten some properties of the hyperbolic model (6.45). For all numerical tests carried out, we take $\alpha_1(N_1) = 0.33$, $D_n = 1$, and $D_{N_1} = 0.001$. For initial conditions, we consider an initial datum for the chemical concentration N_1 and for the flux $n\xi$ which are at rest: $N_1(0) = 0$, and $(n\xi)(0) = 0$.

Concerning the density of cells n, the following specific initial condition is considered:

$$n(0, x, y) = \frac{n_0}{2\pi\sigma^2}\Big(\exp\big(-\frac{(x - x_0)^2 + (y - y_0)^2}{2\sigma^2}\big) + \exp\big(-\frac{(x + x_0)^2 + (y + y_0)^2}{2\sigma^2}\big)\Big),$$

where $n_0 = 0.25$, $(x_0, y_0) = (3\sigma, 3\sigma)$ and $\sigma = 3.10^{-2}$.

The behavior of the model (6.45) in the two-dimensional case is illustrated in Figs. 6.12 and 6.14, where we plot the density of cells at different times ($t = 0.001$, $0.003, 0.005, 0.007, 0.012, 0.015$) (Figures 6.11 and 6.13).

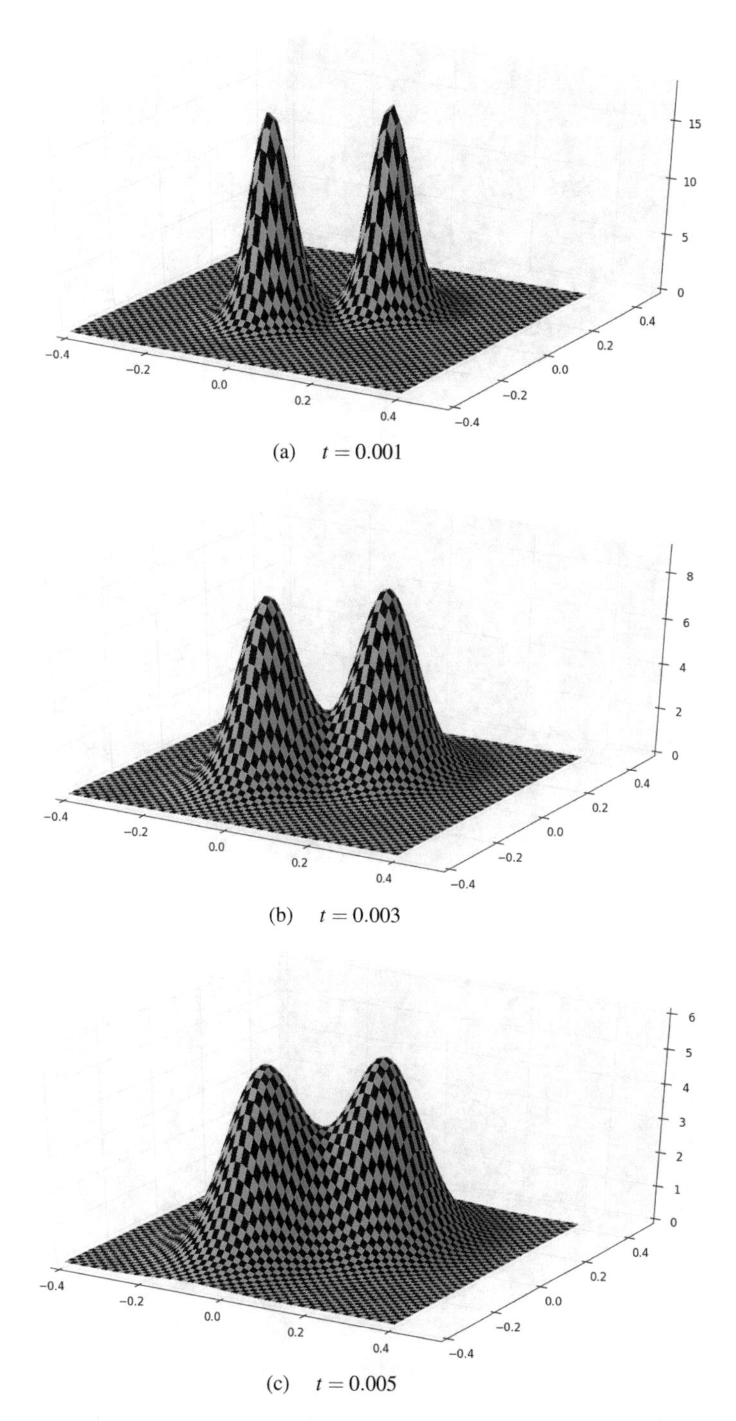

(a) $t = 0.001$

(b) $t = 0.003$

(c) $t = 0.005$

Fig. 6.11 Time dynamics of the cell density $n(t, x)$ in two-dimensional space on a square domain $[-0.4, 0.4] \times [-0.4, 0.4]$ with $s = 100$. Parameter values: $\alpha_1 = 0.33$, $D_n = 1$, $D_{N_1} = 0.001$

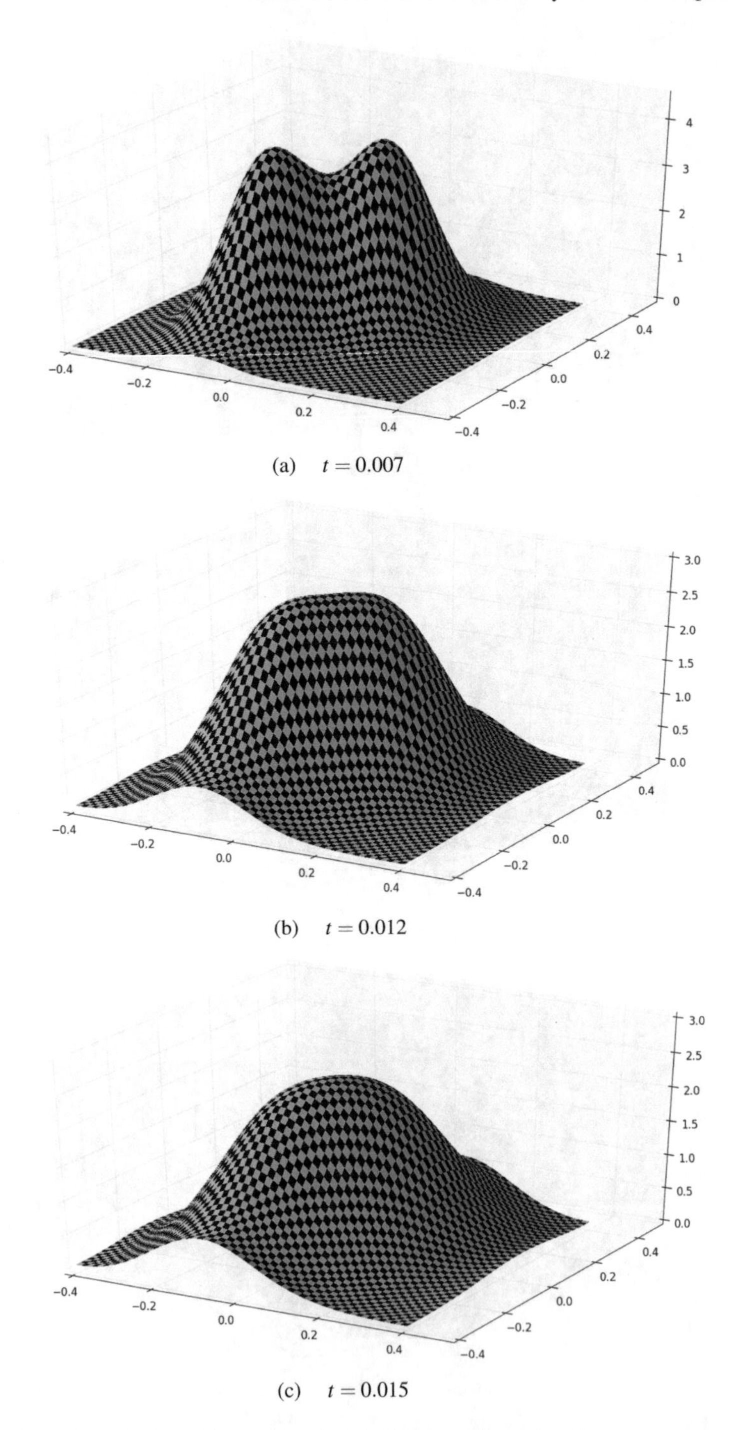

(a) $t = 0.007$

(b) $t = 0.012$

(c) $t = 0.015$

Fig. 6.12 Time dynamics of the cell density $n(t, x)$ in two-dimensional space on a square domain $[-0.4, 0.4] \times [-0.4, 0.4]$ with $s = 100$. Parameter values: $\alpha_1 = 0.33$, $D_n = 1$, $D_{N_1} = 0.001$

Fig. 6.13 Crowd

Fig. 6.14 Swarm

This simple numerical experiment enlightens an approach which is conceptually different from that developed in our book, where simulations are directly applied to the kinetic-type mathematical model. As we have seen, simulations are here obtained for a macroscopic model derived from the underlying description at the microscopic scale. Technically, the result has been obtained by finite volume methods for hyperbolic equations. This approach allows also to identify the blow up of solutions which is a typical feature of chemotaxis models [33].

We need mentioning that this topic has been proposed here without the aim of providing an exhaustive presentation, but simply with the aim of introducing an interesting topic that deserves attention and further developments. The presentation of the literature in the field is postponed to Chapter 7, where it is given in a more general context rather than related to a specific application, while more details on the technicalities involved in this specific applications can be recovered in the already cited paper [195].

6.7 Open Problems and Research Perspectives

The derivation of models has generated a variety of challenging problems still open to the contribution of applied mathematics. Accordingly, we have selected a number of them as possible research perspectives. The first topic is again focused on crowd dynamics, where a number of possible developments are suggested. The most important of them is the modeling of social behaviors in human crowd. Then, a variety of possible applications is presented where technical modifications of the mathematical structures used in this chapter can be used to model large systems of self-propelled particles, namely vehicular traffic and swarms. The next subsections are devoted to a briefly outlining of these perspectives and to provide some suggestions to tackle them.

6.7.1 Social Dynamics in Human Crowds

The present research activity on crowd modeling witnesses a growing interest on the onset and propagation of stress in crowds due to panic conditions that be generated by small groups or even by local overcrowding. Indeed, these studies are motivated by safety problems [210, 211, 245]. Particularly important is the study of evacuation dynamics, where overcrowding is a dangerous event as walkers can get injured (Fig. 6.13). However, this is not the only case study, as additional modeling problems can be generated by more complex social dynamics that have been studied within the framework of socio-psychological studies [3, 244] to be referred also to the role of the features of the venue where walkers move.

The problem of modeling a dynamics where walkers can move across functional subsystems is an important topic. As an example, one can consider a crowd of individuals in a public protest demonstration to support or make opposition to political issues. These individuals can be subdivided into a large group which manifests correctly its position, while a small group is constituted by rioters. The number of the latter can grow in time due to interactions which might persuade the other protesters to join them. Likewise, a small group of security forces might react to provocations out of the standard protocols, but their number might also grow due to interactions, which cause an excess of reactions.

An additional example worth to be studied is the modeling of the dynamics in the presence of leaders in charge of driving the crowd in evacuation. In principle, the leaders improve security conditions based on a well-defined strategy which is not modified by the crowd. Their action is useful to improve safety as the crowd is attracted by leaders and by the rational motion they pursue. If some walkers in the crowd understand and share the leaders' strategy, then these walkers start acting as leaders.

Within the framework proposed in the present chapter, this problem can be dealt with by modifying the transition probability density so as to take into account that interactions not only modify the activity variable, but also allow dynamics across functional subsystems. More specifically, in the case of the first example, due to interactions, particles belonging to the functional subsystem of peaceful protesters can move into the functional subsystem of rioters.

Although the mathematical structure (6.14) is appropriate to take into account the aforementioned hallmarks, transferring the hints into specific models requires an important deal of work to be developed in a proper research program, where the most important improvement would be modeling the complex interactions between social dynamics and mechanical behaviors.

6.7.2 Modeling Vehicular Traffic

This section presents some preliminary rationale to show how the approach developed in the preceding sections can be generalized to model different types of self-propelled particles. Our mind goes rapidly to vehicular traffic and swarm dynamics. The presentation is limited to a critical analysis and concepts leaving the interested reader to their formalization.

The modeling of vehicular traffic has shown, since the pioneering work by Prigogine and Hermann [202], that heterogeneity is an important feature of these systems and kinetic theory methods can be properly developed toward the modeling approach. This book has been already mentioned in Chapter 1, where it has been presented as one of the key scientific contributions that have motivated our book.

Heterogeneity is specifically treated in the paper by Paveri Fontana [198] as drivers are supposed to express a *desired distribution function* that is heterogeneously distributed among them. We wish to stress that this paper provides the first effective modeling approach to heterogeneity of interacting living entities. These two scientific contributions can be regarded as milestones in the kinetic theory approach to self-propelled particles.

The existing literature on modeling vehicular traffic and related analytic and computational problems is reviewed and critically analyzed in the survey [39], while more recent activity is reviewed in [47]. More in detail, we bring to the reader's attention a recent paper [56], where it has been shown that an appropriate modeling of nonlinear interactions leads to the description of some interesting features of vehicular traffic such as:

- Clustering of vehicles with closed velocity;
- Dependence of the said clustering on the quality of the road/environment, namely low quality enhance clustering;
- Ability to reproduce the velocity and flow patterns versus density as observed by empirical data.

This paper motivates further developments such as modeling the following specific dynamics:

1. The effect of the perceived density on the dynamics of the vehicles, for instance to model stop-and-go dynamics;
2. Detailed analysis of the role of heterogeneity of driver–vehicle subsystems on the overall dynamics;
3. Dynamics on multilane flows;
4. The effect of external actions acting on the drivers and specifically on the velocity they impose to vehicles;
5. Modeling the role of boundary conditions as an example induced by the presence of tollgates.

Ultimately, let us stress that models should be used in networks, where the main problem consists in modeling how the vehicle–driver subsystems behave at a junction. A first study on this topic is given in [112] where mechanical dynamics is accounted for and is sharply related to the computational scheme used for the simulations.

6.7.3 Some Reasonings on the Modeling of Swarm Dynamics

Before dealing with modeling issues let us define, referring to [19] and [52], the concept of metric and topological distance. If a metric distance is used, an individual interacts with all other individuals in its neighborhood, namely within a "metric domain". Their number fluctuates dynamically as nearby individuals enter and leave this domain. In [19] and [52], a fixed limit of perception corresponding to their interaction neighborhoods is assumed and should be defined in advance, for example, by defining an interaction range or radius. Outside this neighborhood, particles do not interact with the test particle. In general, metric distance model could provide a good explanation for collective, synchronized motions when the density of individual agents is high.

However, when the density is low, topological domains play a relevant role [19, 80, 152]. In a topological distance, the neighborhood of each individual is dynamical and expands and shrinks continuously. The perception limit is defined by the number of neighbors independently on the absolute distance to the reference individual. This requires to formulate a criterion for identifying the flock-mates. For example, the number of members in a topological domain of birds ranges between 6 and 7 neighbors. Ballerini *et al.* [19] suggested that the topological range of interactions rather than the metric distance should model collective motion of birds. The concept of topological distance becomes important to model collective behaviors.

The modeling of swarms is also motivated by the attraction generated by the observation of the beauty of the shapes formed by birds which appear in the sky during spring and autumn periods (Fig 6.14). Analogous phenomena are observed in other animal systems such as fish schools or cells which aggregate forming particular patterns. The study of swarms involves several challenging problems. For instance, flocking phenomena constitute a typical characteristics of swarms. This problem has been treated in several papers such as [86, 87].

In more detail, the modeling approach presents some conceptual differences with respect to crowd dynamics although some common guidelines are followed:

1. Definition of the main features of behavioral dynamics;
2. Derivation of a general structure suitable to offer the conceptual basis for the derivation of models;
3. Modeling interactions at the microscopic scale to implement the aforesaid structure and derive specific models in unbounded and bounded space domains;
4. Validation of models.

An additional challenging problem is the ability of a swarm to express a collective intelligence [84] or memory to reproduce some patterns related to the environmental conditions [67], which can evolve by learning processes. Indeed, a deep insight on emerging strategies needs to be specifically referred to the type of individuals composing the swarm [97], as well as on the interaction between individual behavior and collective dynamics.

The experimental investigation on swarms differs from that developed in the case of pedestrian crowds. In fact, it is mainly focused on understanding the dynamics (and topology) of the interactions corresponding to different animal species. Furthermore, experiments are addressed to understand emerging behaviors, such as flocking phenomena, break up and aggregation of swarms, or reaction to a predator. The collection of empirical data is generally focused on qualitative, rather than quantitative, aspects. The interested reader can find additional bibliography in the last section of the review paper [39].

Flocks are formed by a large group of autonomous mobile individuals with local interactions between subgroups and a possible group objective or additional property of the swarm. In fact, the aggregation of individuals has a global property, which is not observed in the behavior of an isolated individual, through the interaction of the individual units. Although there are many similarities, the study of each species requires knowledge of the specifics of each group. For example, in studies of self-organization of large flocks of birds, the density is similar between back and front and flock density is independent of flock size [19], while in fish schools high frontal density is observed and also that group density depends on the number of individuals [147].

Bearing all above in mind, let us briefly indicate some important features that should be taken into account in the modeling approach:

- Both long-range interactions with the whole swarm and short-range interactions involving a small number of active particles have to be taken into account. Short-range interactions are, however, non-local and involve a fixed number of neighbors rather than particles within a certain interaction domain.
- The dynamics of interactions can differ in the various zones of the swarm; for instance, from the border to the center of the domain occupied by the swarm. Small stochastic perturbations could be an important feature of the dynamics. In panic conditions, these perturbations can be subject to large deviations.

- The strategic ability and the interaction rules in a swarm can be largely modified when panic conditions occur. Namely, a swarm in normal conditions has a well-defined objective, for instance reaching a certain area zone. However, panic conditions can modify the overall strategy to pursue this objective. Thus, emerging behaviors very different from those observed in normal flow conditions can appear. As an example, fragmentation (de-flocking) phenomena can appear.
- Subdivision into functional subsystems should include different types of components of the swarms, in particular, the presence of leaders and the presence of predators.

Predation and competition are among the key factors determining the natural selection of species. Indeed, predation is one of the most fundamental inter-specific interactions in ecology. Density distribution of prey may have consequences for the capacity of predators to act on their population. The net outcome of predation efficiency, preys population abundance, aggregation, dispersion, panic situations, and prey distribution pattern, in such situations, has not been well studied. Conventional non-spatial models do not allow the implementation of different survival strategies that can lead to the spatial heterogeneity of species distribution. An interesting contribution is delivered by paper [98], where space dynamics is carefully taken into account.

The presence of an external attack of a predator can induce large deviations with respect to the normal behavior of the swarm. This problem has been studied by various authors. Among various others, we refer specifically to [169] and the bibliography therein. Two specific issues seem especially interesting in this field in order to be thoroughly examined:

i) Modifications of the structure;

ii) Modifications of the interaction rules.

The first topic basically means that the system is not closed in the presence of an external action. Therefore, the model needs to be implemented by a term suitable to take into account the interactions between the inner system, namely the swarm, and the predator, where the presence of the predator should be supposed known for each position and velocity values. Therefore, the modeling should consider the dynamics of the predator, its representation in terms of probability distribution, and the interactions between the inner and outer system. This matter has been treated in [52] by modeling the dynamics related to the interaction with the swarm. Supposing that the trajectory of the predator is known, the modeling approach should first describe its presence, namely the distribution g defined over its influence area and subsequently the dynamical response of the swarm. It is worth stressing, toward research perspectives, that the presence of danger can induce relevant modification in the behavior of individuals [152]. Hence the model should take it into account. A possible model has been proposed in [52] simply as a preliminary approach to the problem. However, their formalization into a mathematical framework has been only partially achieved and still needs to be developed.

Finally, let us stress that topological distance and metric distance are two aspects of the same problem. Note that the interdependence between both types of cognition concerning the capacity of decision of an individual: Topological, which requires a criterion for identifying the reference flock-mates of an individual, while metric distance refers to the capacity of the individuals to incorporate signals coming from the interior or exterior of the swarm. At every time step, each individual updates its information, switching between metric domain and topological domain. Two recent approaches to this problem have been developed by [152], where each agent can adjust its interaction range when the density of its neighbors changes. The objective, that should be included in the modeling processes, must be to coordinate both interaction domains in a continuous feedback in order to improve the capacities of the individuals in accordance with those of the group.

Some additional bibliography can now be given focusing on swarm modeling such as [133] on the derivation of kinetic equations from the underlying description at the microscopic scale, and the application of the kinetic theory of active particles to derive new structures within the kinetic theory framework.

6.7.4 Analytic and Computational Problems

While specific applications are (and will be) developed, some challenging analytic problems are brought to the attention of applied mathematicians. Two of them are worth to be mentioned among the various ones. The first one is the qualitative analysis of the initial-boundary value problem, where the results proposed in [29] for a crowd in unbounded domain might be further developed to take into account interactions with walls and obstacles. These interactions are non-local and contribute to modify the local velocity distribution. The derivation of macroscopic models from the underlying description delivered by kinetic models has been developed in [27] for unbounded domains. The study of this challenging problem can be further developed in the presence of obstacles and walls.

Focusing on the qualitative analysis, it is worth mentioning that the study of the initial value problem in unbounded domains for models both of crowds and vehicular traffic does not involve technical difficulties if no constraint is imposed on the local density. In fact, the operator modeling interactions shows to have reasonable regularity (Lipschitz) properties. Therefore, local existence is straightforwardly obtained in suitable Banach spaces [29]. In addition, if the number of active particles does not grow in time, the L_1 norm is preserved and the solution can be prolonged for arbitrarily large times.

However, solutions where the local density is greater than the maximal admissible packing density do not have physical meaning. The key tool to prove existence for large times is that delivered in [27, 29] by implementing the model by realistic interaction rules which take into account that the motion is prevented where the local density reaches the maximal value. However, additional problems arise when the presence of walls and obstacles is taken into account.

An further problem, that has been occasionally mentioned, is the validity of the continuity assumption of the probability distribution over the microscopic state for systems where the number of particles is not large enough to justify this assumption. Such a problem has not yet a fully satisfactory answer in the literature. Arguably, the first criticism on this matter was posed by Daganzo focusing on vehicular traffic [90], while various approaches to overcome it have been proposed to model the dynamics of vehicles or pedestrians as documented, respectively, in papers [56, 111], and therein cited bibliography.

The approach of these papers consists in the discretization of the velocity space, while it can be possibly extended to the full phase space. However, the approach used in the case of vehicular traffic, which is one dimensional, needs further substantial developments to be extended to crowd and swarm dynamics, where a complex geometry is a general feature for a large variety of applications. It is plain that discrete velocity and space methods should be developed consistently with the computational approach. Deterministic methods have been used in [112], while the development of stochastic particles methods for discrete velocity and space needs attention as it can effectively contribute to applications.

Discretization of the velocity space allows a straightforward application of deterministic methods such as those applied in [29], while if the velocity variable is left continuous, particle methods appear to be more efficient.

Chapter 7
On the Search for a Mathematical Theory of Living Systems

7.1 Plan of the Chapter

This chapter proposes some speculations which are freely developed by taking advantage also of an overview and critical analysis of the contents of the preceding chapters. These speculations focus on the fifth key question stated in Chapter 1:

Which are the conceptual paths that lead to
a mathematical theory of living systems?

The answer proposed in the following does not naively claim to be exhaustive. As a matter of fact, the key question poses one of the most challenging problems to be tackled by scientists in this century. However, some preliminary rationale can hopefully contribute to pursue the objective of developing a mathematical theory of living systems.

Indeed, it is a highly challenging objective, but we feel it is worth at least looking for some preliminary results. Our ideas are proposed simply at a qualitative level leaving the mathematical formalization to future steps and research programs.

The following specific topics are selected, among various possible ones, according to the authors' bias and are presented in the next sections.

- Section 7.2 proposes some speculations on the interpretation of *hard* and *soft* sciences. A critical analysis of the existing literature is presented and, out of it, new concepts are proposed as a natural development of the mathematical tools developed in this book.

- Section 7.3 deals with an overview on multiscale methods first by reviewing the literature in the field. Subsequently, some perspective ideas are brought to the attention of a reader interested to study this approach to model important features of living systems.

N. Bellomo et al., *A Quest Towards a Mathematical Theory of Living Systems*,
Modeling and Simulation in Science, Engineering and Technology,
DOI 10.1007/978-3-319-57436-3_7

- Section 7.4 provides a direct answer to the aforementioned fifth key question. The answer starts from a possible definition of mathematical theory of living systems and subsequently proposes some speculations toward a quest to invent such a theory.

7.2 On the Conceptual Distance between Soft and Hard Sciences

Scientists are often involved in the understanding the features of the sciences within the general framework, where their research activity is carried on. One of the possible speculations refers to the conceptual differences between the so-called *hard* and *soft* sciences.

A commonly shared idea is that *hard* sciences generate theories founded on rigorous assumptions supported by experiments, while this approach cannot be applied to *soft* sciences, where a certain amount of heuristic conjectures are needed in their application to real-life problems. According to this drastic division, physics and chemistry can be viewed as hard sciences, while sociology or psychology are soft sciences. Biology is generally put in the middle as in some special cases can take advantage of a field theory.

The scientific community is aware that biology can, in this century, become progressively an hard science supported by mathematical tools and physical–chemical theories. On the other hand, a negative attitude has occasionally appeared to state, not correctly in our opinion, that some soft sciences cannot even be considered a science.

The dispute has been occasionally very strong as documented in an article by Diamond [101] which contains the following quotations:

> *Samuel Huntington, professor of government. Harward* : **The overall correlation between frustration and instability (in 62 countries of the world) was** 0.50.

> *Serge Lang, professor of mathematics, Yale* : **This is utter nonsense. How does Huntington measure things like frustration? Does he have a social frustration meter? I object to the academy's certifying as a science what was merely political opinions.**

> *Other scholars, commenting on Lang's attack* : **What does it say about Lang's scientific standards that he base his case on twenty years old gossip? . . . a bizarre vendetta.**

Nowadays, mathematicians appear to be able to look with positivism, rather than negativism, to soft sciences and show the courage of quantifying *soft variables* [221].

Mathematics is somehow different as it can be viewed as a formalized architecture based on a number of axioms. One might even be induced to think that it is a totally abstract speculative science. However, this observation, which does not look forward, fails not only when we look at the origin of mathematics certainly motivated by a need to provide a formal interpretation of physical reality, but also mainly when we look at the interactions between physics and mathematics that have been essential to the development of physical theories.

Now the key point is understanding how the interactions between mathematics and soft sciences can be organized.

The contents of this book support the idea that interactions can be put in a rationale that can contribute to the progress of both soft sciences and mathematics that can receive from the aforementioned interaction hints to develop new mathematical tools and even theories. Hence, the aforementioned negative attitude should be put apart.

A first constructive observation is that a hierarchy from hard to soft should be introduced according to the different levels of possible invention of field theory to support each science. Actually, a continuous classification should replace drastic separations and even a hierarchy.

This rationale appears to be interesting, but it should also take into account the interdisciplinary characterization of modern research activity.

As a possible example, we have seen in Chapter 5 that the complex interplay between economy, sociology, and human behaviors can enlighten research programs. Similarly, Chapter 6 has shown that the dynamics of crowds and swarms is influenced by social dynamics and interactions.

Our wish consists in taking to the table another guest, however always present, namely mathematical sciences. In fact, the classification of all sciences depends also on their possible interactions with mathematics.

A consequence of all above reasonings is that rather than speculating on hard and soft sciences, we suggest to look at the inert and living matter by a sharp interpretation of the role of *life* in each specific system rather than in an overall discipline. Indeed, the role of *life* can be weaker or stronger in each system compared with the others.

Therefore, let us return to the interaction between mathematics and soft sciences and let us substitute it by

Interaction between mathematics and living systems.

This idea is consistent with what was expressed by Hartwell [137] as reported in Chapter 1, where he stated that the mathematics of the inert matter cannot be straightforwardly applied to living systems. Furthermore, when this is done, mathematics generally fails [243].

An additional topic to complete our reasonings is that living systems are always evolutionary and, due to a struggle with the external environment, generally for survival, they can mutate and undergo a Darwinian-type selection. Accordingly, the fundamental ideas by Mayr [173, 174] and coworkers [102, 103] are a precious gift to science and mathematics.

Let us now observe that this section has not yet given an answer to the fifth key question, while the conceptual basis to do it has simply been just introduced. Then, waiting for a constructive reply, we wish stressing that mathematical sciences receive a great deal motivations from the sciences of the living matter to develop new mathematical tools and possibly theories.

Indeed, it is not an easy task [150, 172, 205]. Perhaps it is even too difficult; however, we insists in stating that it is worth trying. In fact, pursuing this challenging objective generally brings the "gift" to generate interesting mathematical problems which, at least, contribute to the progress of mathematical sciences.

7.3 Multiscale Issues Toward a Mathematical Theory

The mathematical tools proposed in this book have been applied to the modeling of large systems of interacting entities called *active particles*, whose microscopic state includes an *activity variable*. The mathematical approach provides the dynamics of the probability distribution over the said state. Therefore, macroscopic quantities are obtained by weighted moments of the probability distribution that is the dependent variable in the differential system. It is worth stressing that the solution to mathematical problems provides a richer description as it gives the shape of the distribution, of active particles at each time and each space location.

By multiscale problem, we intend the search of the analytic links bridging the dynamics through the three different scales: microscopic, mesoscopic, and macroscopic, as represented in Fig. 7.1.

The mathematical structures, proposed in this book, operate at the *microscopic scale* by kinetic theory methods and theoretical tools of stochastic game theory. The recent literature has shown that the micro–macro passage can be obtained by

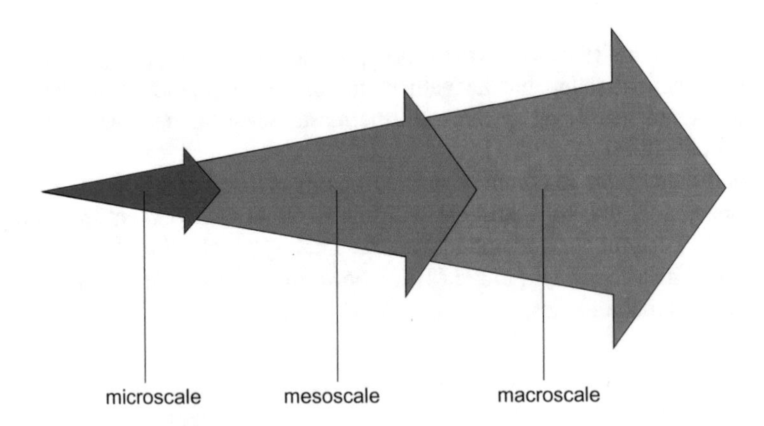

Fig. 7.1 From the microscale to the macroscale

asymptotic limits for specific classes of equations written in an appropriate dimensionless form such that a dimensionless parameter appears corresponding to the physical distance between active particles.

Asymptotic limits are obtained by letting to zero, under appropriate scaling assumptions, this parameter. Then, under suitable convergence proofs, models at the macroscopic scale can be derived. This approach was introduced in biology by Othmer, Dunbar, and Alt [193], where the main idea consists in perturbing the transport equation by a velocity jump process, which appears appropriate to model the velocity dynamics of cells modeled as living particles.

This method, which leads to hyperbolic transport equation, has been subsequently developed by various authors by different approaches and scaling corresponding both to parabolic and hyperbolic. We can cite among others and without any claim of completeness the following papers [7, 28, 30, 31, 55, 113, 153, 194, 195] that contributed to improve the approach and developed various applications in biology.

More recent applications have focused on vehicular traffic [32] and crowd dynamics [27], while paper [33] has shown how different approaches can lead to the derivation of chemotaxis models. The book by Banasiak and Lachowicz [20] is an important source of additional methods and bibliography.

An important output of this literature is that the micro–macro derivation generates models that can be different from that of the direct phenomenological approach consisting in the use of conservation equations closed by heuristic models of material behaviors. In some cases, these new models show interesting qualitative features consistent with physical reality, that are not shown by purely phenomenological models.

It is worth mentioning that the sixth Hilbert problem [151] is addressed to the search for a unified mathematical approach to physical theories. This also means looking for the links between different scales when a given system is described by mathematical equations. As an example, the equations of hydrodynamics should be obtained from the underlying description at the microscopic scale.

Focusing on this specific problem, David Hilbert studied the derivation of equations of hydrodynamics starting from the statistical description of particles delivered by the Boltzmann equation. He proposed an expansion of the distribution function in terms of a small parameter modeling intermolecular distances. Equating terms of the same power of the parameter and exploiting that the zero-order solution is given by Maxwell distribution at equilibrium leads to the formal derivation of hydrodynamic equations. This approach replaces the heuristic derivation obtained by conservation equations closed by phenomenological models of material behaviors.

Hilbert's celebrated models have generated a broad literature reviewed in Chapters 3 and 11 of [82]. The scientific community agrees on the relevance of the contribution by Laure Saint-Raymond presented in the Lectures Notes [213], where several conceptual issues received the appropriate answer within the rigorous framework of mathematical sciences.

After this rapid outlook to classical methods, we can return to the micro–macro derivation of large systems of active (living) particles and mention that the Hilbert's method has been further developed by two recent papers [54, 74] to derive macroscopic equations from the underlying description delivered by the kinetic theory of multicellular systems. In more detail, binary mixtures have been studied in [74] focusing on parabolic–parabolic, parabolic–hyperbolic, hyperbolic–hyperbolic time–space scaling. The approach has been applied to the derivation of possible variations of the celebrated Keller–Segel model.

These results appear to be definitely interesting and open a window over future developments even for sophisticated systems which might include mutations and selection related to proliferative–destructive events. Hence, a fascinating research field is opened by [74]. However, although we can conclude that it is now well understood how the general structure proposed in Chapter 3 leads to the micro–macro derivation.

Unfortunately, far less understood is the bridge between the submicroscopic and the microscopic scales. This is an important topic as the dynamics of active particles depends also on the dynamics at the lower scale. This complex relationship has been not yet put in a mathematical framework. A mathematical conjecture was proposed in [38] based on the idea that structures analogous to that used at the microscopic scale can be used also at the lower scale. In biology, the scale of genes is below that of cells. If the dynamics can be computed by models at the lowest scale, then the dynamical properties of active particles can also be modeled.

Definitely, a success in tackling this problem can contribute to the derivation of a mathematical theory of living systems as it is further discussed in the last subsection.

7.4 Waiting for a Mathematical Theory of Living Systems

The main conceptual problem, which has motivated this book, is the search for a mathematical theory of living systems. This challenging objective cannot, at least at present, be developed within a very general approach. However, some very preliminary achievements can be obtained after having specialized the search to very specific fields.

As we have seen the first step is the derivation of a general mathematical structure suitable to capture the main complexity features of living systems. Namely, the ability to express a strategy/purpose, heterogeneity, nonlinear interactions, learning ability, and Darwinian-type mutations, and selection. Heterogeneity includes, as we have seen, the subdivision of the overall system into functional subsystems such that each of them express a different strategy.

This structure offers the framework for the derivation of models. Various examples have been presented in Chapters 5 and 6. The structure can be viewed as a mathematical theory as it consists in a new class of equations, whose qualitative properties have been only partially understood, while the analysis is still object of research activity. In addition, computational methods have been specifically developed to obtain quantitative results for problems generated by the application of models to real physical situations. Computation accounts for the stochastic feature of living systems and hence of the mathematical approach to modeling.

An additional step toward the derivation of a mathematical theory of living systems is the multiscale analysis outlined in Subsection 7.3. However, the derivation of the said theory needs theoretical tools to describe by equations the complex interactions of living entities. Indeed, this is common to all physical theories, where, as it is known, the overall dynamics at a certain observation scale depends on the dynamics, or at least properties, at the lower scale. As an example, the dynamics of classical particles needs modeling their interactions, which depend on their physical properties and are based on the application of principles of Newtonian mechanics.

A phenomenological interpretation of reality can lead to modeling interactions. Their implementation in the aforementioned structure leads to the derivation of models. If the process is successful, then a valid model has been derived. However, it is a just model and not a mathematical theory.

The theory we are dreaming about is still far away as we do not have yet theoretical tools and experiments to support the modeling of interactions. Definitely a deeper understanding of the link between the lowest scales and that of the active particles can contribute to this challenging objective. The present state of the art indicates that we are still far away from it. However, some possible perspectives appear in the field of biology, where the important recent achievements at the molecular scales of genes permit to look with some optimism at this challenging objective. More far away is the target for sciences where "life" is more present. Indeed:

Jared Diamond : **Sciences are often harder than hard sciences** [101].

Let us now start from this general statement and look at it with the eyes of a mathematician. The overall historical development of mathematical sciences has generated a sequence of fascinating problems, some of them abstract, some related to applications.

Abstract problems have been (and are) a continuous challenge for mathematicians, let us think about some mysterious properties of prime numbers or to Fermat conjectures. When one of these milestone problems is solved, the whole scientific community, namely not only mathematicians, looks at such results with great enthusiasm and admiration.

Mathematics of real world looks at problems in a different way. Namely, when a new interpretation of physical reality is produced, mathematicians work at it and, in various cases, a new mathematical theory is produced. This theory is a gift for mathematicians who operate toward a deeper understanding of the properties of such a theory. It does not really matter if the study carry them far from the physical reality generated by the theory as it can be an interesting mathematical speculation hopefully worth for future applications.

Our opinion is that this way of thinking should not be negatively interpreted. In fact, these speculations can lead to improvements of the theory as well as its possible generalizations to different field of applications. Soft sciences offer a precious opportunity to develop a new mathematical theory. Hence, they offer a challenging opportunity to mathematicians.

Returning now to the aims of this book, our closure is as follows:

The mathematical structure presented in this book and its applications have generated a variety of challenging analytic and computational problems. New ones will arguably appear in a next future. Can such a structure be considered as a first step toward the invention of a mathematical theory of living systems?

Hence, our closure appears as a new key question. Although we trust that the answer will be given in the next years, we believe that we can state the following:

Yes, it is a first step.

References

1. Acemoglu, D., Bimpikis, K., Ozdaglar, A.: Dynamics of information exchange in endogenous social networks. National Bureau of Economic Research, Paper 16410 (2010)
2. Agnelli, J.P., Colasuonno, F., Knopoff, D.: A kinetic theory approach to the dynamics of crowd evacuation from bounded domains. Math. Models Methods Appl. Sci. **25**, 109–129 (2015)
3. Aguirre, B., Wenger, D., Vigo, G.: A test of the emergent norm theory of collective behavior. Soc. Forum **13**, 301–311 (1998)
4. Ajmone Marsan, G., Bellomo, N., Egidi, M.: Towards a mathematical theory of complex socio-economical systems by functional subsystems representation. Kinet. Relat. Models **1**(2), 249–278 (2008)
5. Ajmone Marsan, G., Bellomo, N., Gibelli, L.: Stochastic evolutionary differential games toward a system theory of behavioral social dynamics. Math. Models Methods Appl. Sci. **26**(6), 1051–1093 (2016)
6. Ajmone Marsan, G., Bellomo, N., Tosin, A.: Complex System and Society: Modeling and Simulation. Springer, New York (2013)
7. Alber, M., Chen, N., Lushnikov, P., Newman, S.: Continuous macroscopic limit of a discrete stochastic model for interaction of living cells. Phys. Rev. Lett. **99**, 168102 (2007)
8. Albert, R., Barabási, A.L.: Statistical mechanics of complex networks. Rev. Mod. Phys. **74**, 47–97 (2002)
9. Alekseenko, A., Josyula, E.: Deterministic solution of the Boltzmann equation using discontinuous galerkin discretization in velocity space. J. Comput. Phys. **272**, 170–188 (2014)
10. Aristov, V.V.: Direct Methods for Solving the Boltzmann Equation and Study of Nonequilibrium Flows. Springer, New York (2001)
11. Aristov, V.V., Tcheremissine, F.G.: The conservative splitting method for the solution of a Boltzmann. U.S.S.R Comput. Math. Phys. **20**, 208–225 (1980)
12. Arlotti, L., Bellomo, N., De Angelis, E.: Generalized kinetic (Boltzmann) models: mathematical structures and applications. Math. Models Methods Appl. Sci. **12**(2), 567–591 (2002)
13. Arlotti, L., Bellomo, N., Lachowicz, M., Polewczak, J., Walus, W.: Lectures Notes on the Mathematical Theory of the Boltzmann Equation. World Scientific, Singapore (1995)
14. Arlotti, L., De Angelis, E., Fermo, L., Lachowicz, M., Bellomo, N.: On a class of integro-differential equations modeling complex systems with nonlinear interactions. Appl. Math. Lett. **25**, 490–495 (2012)
15. Arthur, W., Durlauf, S., Lane, D. (eds.): The Economy as an Evolving Complex System II. Studies in the Sciences of Complexity, vol. XXVII. Addison-Wesley, Reading (1997)
16. Axelrod, R.M.: The Complexity of Cooperation: Agent-based Models of Competition and Collaboration. Princeton University Press, Princeton (1997)

N. Bellomo et al., *A Quest Towards a Mathematical Theory of Living Systems*,
Modeling and Simulation in Science, Engineering and Technology,
DOI 10.1007/978-3-319-57436-3

17. Babovsky, H., Illner, R.: A convergence proof for Naubu's simulation method for the full Boltzmann equation. Math. Methods Appl. Sci. **8**, 223–233 (1986)
18. Ball, P.: Why Society is a Complex Matter. Springer, Heidelberg (2012)
19. Ballerini, M., Cabibbo, N., Candelier, R., Cavagna, A., Cisbani, E., Giardina, I., Lecomte, V., Orlandi, A., Parisi, G., Procaccini, A., Viale, M., Zdravkovic, V.: Interaction ruling animal collective behavior depends on topological rather than metric distance: evidence from a field study. Proc. Natl. Acad. Sci. **105**, 1232–1237 (2008)
20. Banasiak, J., Lachowicz, M.: Methods of Small Parameter in Mathematical Biology. Birkhäuser, Boston (2014)
21. Barabási, A.L.: The Science of Networks. Perseus, Cambridge (2012)
22. Barabási, A.L., Albert, R.: Emergence of scaling in random networks. Science **286**(5439), 509–512 (1999)
23. Bardi, M., Cesaroni, A., Ghilli, D.: Large deviations for some fast stochastic volatility models by viscosity methods. Discret. Contin. Dyn. Syst. Ser. A **32**, 3965–3988 (2015)
24. Barrat, A., Barthélemy, M., Vespignani, A.: The Structure and Dynamics of Networks. Princeton University Press, Princeton (2006)
25. Bastolla, U., Fortuna, M., Pascual-García, A., Ferrera, A., Luque, B., Bascompte, J.: The architecture of mutualistic networks minimizes competition and increases biodiversity. Nature **458**, 1018–1020 (2009)
26. Bellomo, N.: Modeling Complex Living Systems - A Kinetic Theory and Stochastic Game Approach. Birkhäuser, Boston (2008)
27. Bellomo, N., Bellouquid, A.: On multiscale models of pedestrian crowds from mesoscopic to macroscopic. Commun. Math. Sci. **13**, 1649–1664 (2015)
28. Bellomo, N., Bellouquid, A., Chouhad, N.: From a multiscale derivation of nonlinear cross diffusion models to Keller-Segel models in a Navier-Stokes fluid. Math. Models Methods Appl. Sci. **26**, 2041–2069 (2016)
29. Bellomo, N., Bellouquid, A., Knopoff, D.: From the micro-scale to collective crowd dynamics. Multiscale Model. Simul. **11**, 943–963 (2013)
30. Bellomo, N., Bellouquid, A., Nieto, J., Soler, J.: Multicellular growing systems: hyperbolic limits towards macroscopic description. Math. Models Methods Appl. Sci. **17**, 1675–1693 (2007)
31. Bellomo, N., Bellouquid, A., Nieto, J., Soler, J.: Multiscale biological tissue models and flux-limited chemotaxis from binary mixtures of multicellular growing systems. Math. Models Methods Appl. Sci. **20**, 1179–1207 (2010)
32. Bellomo, N., Bellouquid, A., Nieto, J., Soler, J.: On the multi scale modeling of vehicular traffic: from kinetic to hydrodynamics. Discret. Contin. Dyn. Syst. Ser. B **19**, 1869–1888 (2014)
33. Bellomo, N., Bellouquid, A., Tao, Y., Winkler, M.: Toward a mathematical theory of Keller-Segel models of pattern formation in biological tissues. Math. Models Methods Appl. Sci. **25**, 1663–1763 (2015)
34. Bellomo, N., Carbonaro, B.: On the complexity of multiple interactions with additional reasonings about Kate. Jules and Jim. Math. Comput. Model. **47**, 168–177 (2008)
35. Bellomo, N., Colasuonno, F., Knopoff, D., Soler, J.: From a systems theory of sociology to modeling the onset and evolution of criminality. Netw. Heterog. Media **10**, 421–441 (2015)
36. Bellomo, N., Coscia, V.: Sources of nonlinearity in the kinetic theory of active particles with focus on the formation of political opinions. AMS Ser. Contemp. Math. **594**, 99–114 (2015)
37. Bellomo, N., Degond, P., Tadmor, E.: Active Particles - Volume 1 - Theory, Models, Applications. Birkhauser, New York (2017)
38. Bellomo, N., Delitala, M.: On the coupling of higher and lower scales using the mathematical theory of active particles. Appl. Math. Lett. **22**, 646–650 (2009)
39. Bellomo, N., Dogbè, C.: On the modelling of traffic and crowds - a survey of models, speculations, and perspectives. SIAM Rev. **53**, 409–463 (2011)
40. Bellomo, N., Elaiw, A., Althiabi, A., Alghamdi, A.: On the interplay between mathematics and biology hallmarks toward a new systems biology. Phys. Life Rev. **12**, 44–64 (2015)

41. Bellomo, N., Forni, G.: Dynamics of tumor interaction with the host immune system. Math. Comput. Model. **20**, 107–122 (1994)
42. Bellomo, N., Gatignol, R.: Lecture Notes on the Discretization of the Boltzmann Equation. World Scientific, Singapore (2003)
43. Bellomo, N., Gibelli, L.: Toward a behavioral-social dynamics of pedestrian crowds. Math. Models Methods Appl. Sci. **25**, 2417–2437 (2015)
44. Bellomo, N., Gibelli, L.: Behavioral crowds: modeling and Monte Carlo simulations toward validation. Comput. Fluids **141**, 13–21 (2016)
45. Bellomo, N., Gustafsson, T.: The discrete Boltzmann equation: a review of the mathematical aspects of the initial and initial-boundary value problems. Rev. Math. Phys. **3**, 213–255 (1991)
46. Bellomo, N., Herrero, M.A., Tosin, A.: On the dynamics of social conflicts: looking for the Black Swan. Kinet. Relat. Models **6**, 459–479 (2013)
47. Bellomo, N., Knopoff, D., Soler, J.: On the difficult interplay between life, "complexity", and mathematical sciences. Math. Models Methods Appl. Sci. **23**, 1861–1913 (2013)
48. Bellomo, N., Lachowicz, M., Polewczak, J., Toscani, G.: Mathematical Topics in Nonlinear Kinetic Theory II: The Enskog Equation. World Scientific, Singapore (1991)
49. Bellomo, N., Palczewski, A., Toscani, G.: Mathematical Topics in Nonlinear Kinetic Theory. World Scientific, Singapore (1988)
50. Bellomo, N., Piccoli, B., Tosin, A.: Modeling crowd dynamics from a complex system viewpoint. Math. Models Methods Appl. Sci. **22**, 1230004 (2012)
51. Bellomo, N., Pulvirenti, M. (eds.): Modeling in Applied Sciences - A Kinetic Theory Approach. Birkhäuser, Boston (2000)
52. Bellomo, N., Soler, J.: On the mathematical theory of the dynamics of swarms viewed as complex systems. Math. Models Methods Appl. Sci. **22**, 1140006 (2012)
53. Bellomo, N., Toscani, T.: On the Cauchy problem for the nonlinear Boltzmann equation: global existence, uniqueness, and asymptotic behaviour. J. Math. Phys. **26**, 334–338 (1985)
54. Bellouquid, A., Chouhad, N.: Kinetic models for chemotaxis toward diffusive limits: asymptotic analysis. Math. Methods Appl. Sci. **39**, 3136–3151 (2016)
55. Bellouquid, A., De Angelis, E.: From kinetic models of multicellular growing systems to macroscopic biological tissue models. Nonlinear Anal. RWA **12**, 1101–1122 (2011)
56. Bellouquid, A., De Angelis, E., Fermo, L.: Towards the modeling of vehicular traffic as a complex system: a kinetic theory approach. Math. Models Methods Appl. Sci. **22**, 1140003 (2012)
57. Bellouquid, A., De Angelis, E., Knopoff, D.: From the modeling of the immune hallmarks of cancer to a black swan in biology. Math. Models Methods Appl. Sci. **23**, 949–978 (2013)
58. Bellouquid, A., Delitala, M.: Mathematical methods and tools of kinetic theory towards modelling complex biological systems. Math. Models Methods Appl. Sci. **15**, 1639–1666 (2005)
59. Berenji, B., Chou, T., D'Orsogna, M.: Recidivism and rehabilitation of criminal offenders: a carrot and stick evolutionary games. PLOS ONE **9**, 885831 (2014)
60. Bertotti, M.L., Delitala, M.: From discrete kinetic and stochastic game theory to modelling complex systems in applied sciences. Math. Models Methods Appl. Sci. **14**(7), 1061–1084 (2004)
61. Bertotti, M.L., Delitala, M.: Conservation laws and asymptotic behavior of a model of social dynamics. Nonlinear Anal. Real World Appl. **9**(1), 183–196 (2008)
62. Bertotti, M.L., Modanese, G.: From microscopic taxation and redistribution models to macroscopic income distributions. Physica A **390**, 3782–3793 (2011)
63. Besley, T., Persson, T., Sturm, D.: Political competition, policy and growth: theory and evidence from the US. Rev. Econ. Stud. **77**, 1329–1352 (2010)
64. Bird, G.: Molecular Gas Dynamics and the Direct Simulation of Gas Flows. Oxford University Press, Oxford (1994)
65. Bird, G.A.: Approach to translational equilibrium in a rigid sphere gas. Phys. Fluids **6**, 1518–1519 (1963)
66. Bisin, A., Verdier, T.: The economics of cultural transmission and the dynamics of preferences. J. Econ. Theory **97**, 298–319 (2001)

67. Bonabeau, E., Dorigo, M., Theraulaz, G.: Swarm Intelligence: From Natural to Artificial Systems. Oxford University Press, Oxford (1999)
68. Bonacich, P., Lu, P.: Introduction to Mathematical Sociology. Princeton University Press, Princeton (2012)
69. Bressan, A.: Bifurcation analysis of a non-cooperative differential game with one weak player. J. Differ. Equ. **248**(6), 1297–1314 (2010)
70. Bressan, A.: Noncooperative differential games. A tutorial, Lecture Notes for a Summer Course. http://descartes.math.psu.edu/bressan/PSPDF/game-lnew.pdf (2010)
71. Bressan, A., Shen, W.: Semi-cooperative strategies for differential games. Int. J. Game Theory **32**(4), 561–593 (2004)
72. Bürger, R.: The Mathematical Theory of Selection. Recombination and Mutation. Wiley, London (2000)
73. Bürger, R., Willensdorfer, M., Nowak, M.: Why are phenotypic mutation rates much higher than genotypic mutation rates? Genetics **172**, 197–206 (2006)
74. Burini, D., Chouhad, N.: An Hilbert method toward multiscale analysis: from kinetic to macroscopic models for active particles. Math. Models Methods Appl. Sci. **27**(6), 1327–1353 (2017)
75. Burini, D., De Lillo, S., Gibelli, L.: Stochastic differential "nonlinear" games modeling collective learning dynamics. Phys. Life Rev. **16**, 123–139 (2016)
76. Camerer, C.: Behavioral Game Theory: Experiments in Strategic Interaction. Princeton University Press, Princeton (2003)
77. Carrillo, J., Fornasier, M., Rosado, J., Toscani, G.: Asymptotic flocking dynamics for the kinetic Cucker-Smale model. SIAM J. Math. Anal. **42**, 218–236 (2010)
78. Castellano, C., Fortunato, S., Loreto, V.: Statistical physics of social dynamics. Rev. Modern Phys. **81**, 591–646 (2009)
79. Cattani, C., Ciancio, A.: Hybrid two scales mathematical tools for active particles modelling complex systems with learning hiding dynamics. Math. Models Methods Appl. Sci. **17**, 171–187 (2007)
80. Cavagna, A., Cimarelli, A., Giardina, I., Parisi, G., Santagati, R., Stefanini, F., Viale, M.: Scale-free correlation in starling flocks. Proc. Natl. Acad. Sci. **107**, 11865–11870 (2010)
81. Cavallo, F., De Giovanni, C., Nanni, P., Forni, G., Lollini, P.L.: 2011: the immune hallmarks of cancer. Cancer Immunol. **60**, 319–326 (2011)
82. Cercignani, C., Illner, R., Pulvirenti, M.: The Mathematical Theory of Diluted Gas. Springer, Heidelberg (1993)
83. Cooper, E.: Evolution of immune system from self/not self to danger to artificial immune system. Phys. Life Rev. **7**, 55–78 (2010)
84. Couzin, I.: Collective minds. Nature **445**, 715 (2007)
85. Cristiani, E., Piccoli, B., Tosin, A.: Multiscale Modeling of Pedestrian Dynamics. Springer, Berlin (2014)
86. Cucker, F., Dong, J.: On the critical exponent for flocks under hierarchical leadership. Math. Models Methods Appl. Sci. **19**, 1391–1404 (2009)
87. Cucker, F., Smale, S.: Emergent behavior in flocks. IEEE Trans. Autom. Control **52**, 853–862 (2007)
88. Daamen, W., Hoogedorn, S.: Experimental research of pedestrian walking behavior. In: TRB Annual Meeting CD-ROM (2006)
89. D'Acci, L.: Mathematize urbes by humanizing them. cities as isobenefit landscapes: psycho-economical distances and personal isobenefit lines. Landsc. Urban Plan. **139**, 63–81 (2015)
90. Daganzo, C.: Requiem for second-order fluid approximation of traffic flow. Transp. Res. B **29**, 277–286 (1995)
91. De Angelis, E.: On the mathematical theory of post-darwinian mutations, selection, and evolution. Math. Models Methods Appl. Sci. **24**(1), 2723–2742 (2014)
92. De Lillo, S., Delitala, M., Salvatori, M.: Modelling epidemics and virus mutations by methods of the mathematical kinetic theory for active particles. Math. Models Methods Appl. Sci. **19**, 1405–1426 (2009)

93. De Montis, A., Barthélemy, M., Chessa, A., Vespignani, A.: The structure of inter-urban traffic: a weighted network analysis. Environ. Plan. B **34**, 905–924 (2007)
94. Degond, P., Appert-Rolland, C., Moussaïd, M., Pettré, J., Theraulaz, G.: A hierarchy of heuristic-based models of crowd dynamics. J. Stat. Phys. **152**, 1033–1068 (2013)
95. Delitala, M., Lorenzi, T.: A mathematical model for value estimate with public information and herding. Kinet. Relat. Models **7**, 29–44 (2014)
96. Delitala, M., Pucci, P., Salvatori, M.: From methods of the mathematical kinetic theory for active particles to modelling virus mutations. Math. Models Methods Appl. Sci. **21**, 843–870 (2011)
97. Deutsch, A., Dormann, S.: Cellular Automaton Modeling of Biological Pattern Formation: Characterization, Applications, and Analysis. Birkhäuser, Boston (2005)
98. Di Francesco, M., Fagioli, S.: A nonlocal swarm model for predators-prey interactions. Math. Models Methods Appl. Sci. **26**, 319–355 (2016)
99. Di Perna, L., Lions, P.: On the cauchy problem for the Boltzmann equation. global existence and weak stability results. Ann. Math. **130**, 1189–1214 (1990)
100. Di Perna, L., Lions, P.: Arch. Ration. Mech. Anal. **114**, 47–55 (1991)
101. Diamond, J.: Soft sciences are often harder than hard sciences. Discover August, pp. 34–39 (1987)
102. Diamond, J.: Obituary: Ernst Mayr (1904–2005). Nature **433**, 700–701 (2005)
103. Diamond, J.: Linguistics: deep relationships between languages. Nature **476**, 291–292 (2011)
104. Diekmann, O., Heesterbeck, J.: Mathematical Epidemiology of Infectious Diseases. Wiley, London (2000)
105. Diekmann, O., Heesterbeek, H., Britton, T.: Mathematical Tools for Understanding Infectious Disease Dynamics. Princeton University Press, Princeton (2013)
106. Dimarco, G., Pareschi, L.: Numerical methods for kinetic equations. Acta Numer. **23**, 369–520 (2014)
107. Dolfin, M., Lachowicz, M.: Modeling altruism and selfishness in welfare dynamics: the role of nonlinear interactions. Math. Models Methods Appl. Sci. **24**, 2361–2381 (2014)
108. Epstein, J.M., Axtell, R.: Growing Artificial Societies: Social Science from The Bottom Up. The MIT Press, Cambridge (1996)
109. Fajnzlber, P., Lederman, D., Loayza, N.: Inequality and violent crime. J. Law Econ. **45**, 1–40 (2002)
110. Felson, M.: What every mathematician should know about modelling crime. Eur. J. Appl. Math. **21**, 275–281 (2010)
111. Fermo, L., Tosin, A.: Fundamental diagrams for kinetic equations of traffic flow. Discret. Contin. Dyn. Syst. Ser. S **7**(3), 449–462 (2014)
112. Fermo, L., Tosin, A.: A fully-discrete-state kinetic theory approach to traffic flow on road networks. Math. Models Methods Appl. Sci. **25**(3), 423–461 (2015)
113. Filbet, F., Laurençot, P., Perthame, B.: Derivation of hyperbolic models for chemosensitive movement. J. Math. Biol. **50**, 189–207 (2005)
114. Filbet, F., Mouhot, C., Pareschi, L.: Solving the Boltzmann equation in n log2 n. SIAM J. Sci. Comput. **28**(3), 1029–1053 (2006)
115. Frank, S.: Dyn. Princeton University Press, Princeton, Cancer Inherit. Evol (2007)
116. Frezzotti, A.: Numerical study of the strong evaluation of a binary mixture. Fluid Dyn. Res. **8**, 175–187 (1991)
117. Frezzotti, A., Ghiroldi, G.P., Gibelli, L., Bonucci, A.: DSMC simulation of rarefied gas mixtures flows driven by arrays of absorbing plates. Vacuum **103**, 57–67 (2014)
118. Frezzotti, A., Ghiroldi, G.P., Gibelli, L.: Rarefied gas mixtures flows driven by surface absorption. Vacuum **86**(11), 1731–1738 (2012)
119. Frezzotti, A., Ghiroldi, G.P., Gibelli, L.: Solving model kinetic equations on GPUs. Comput. Fluids **50**, 136–146 (2011)
120. Frezzotti, A., Ghiroldi, G.P., Gibelli, L.: Solving the Boltzmann equation on GPUs. Comput. Phys. Commun. **182**, 2445–2453 (2011)
121. Galam, S.: Sociophysics. Springer, New York (2012)

122. Gamba, I.M., Tharkabhushanam, S.: Spectral-lagrangian methods for collisional models of non-equilibrium statistical states. J. Computational Physics **228**, 2012–2036 (2009)
123. Garlaschelli, D., Loffredo, M.: Effect of network topology on wealth distribution. J. Phys. A Math. Theor. **41**, 2240018 (2008)
124. Garlaschelli, D., Loffredo, M.: Maximum likelihood: extracting unbiased information from complex networks. Phys. Rev. E **78**, 015101 (2008)
125. Gatignol, R.: Theorie Cinétique des Gaz a Repartition Discréte de Vitesses. Lecture Notes in Physics, vol. 36. Springer, Berlin (1975)
126. Ghiroldi, G., Gibelli, L.: A direct method for the Boltzmann equation based on a pseudo-spectral velocity space discretization. J. Comput. Phys. **258**, 568–584 (2014)
127. Gibelli, L., Elaiw, A., Althiabi, A., Alghamdi, A.M.: Heterogeneous population dynamics of active particles: mutation selection, and entropy dynamics. Math. Models Methods Appl. Sci. **27**(4), 617–640 (2017)
128. Gintis, H.: Beyond homo economicus: evidence from experimental economics. Ecol. Econ. **35**, 311–322 (2000)
129. Gintis, H.: The Bounds of Reason: Game Theory and the Unification of the Behavioral Sciences. Princeton University Press, Princeton (2009)
130. Gintis, H.: Game Theory Evolving: A Problem-Centered Introduction to Modeling Strategic Interaction. Princeton University Press, Princeton (2009)
131. Glassey, R.: The Cauchy Problem in Kinetic Theory. SIAM Publisher, New York (1996)
132. Gromov, M.: In a search for a structure, part 1: On entropy. Preprint. http://www.ihes.fr/~gromov/
133. Ha, S.-Y., Tadmor, E.: From particle to kinetic and hydrodynamic description of flocking. Kinet. Relat. Models **1**, 415–435 (2008)
134. Hanahan, D., Weinberg, R.A.: Hallmarks of cancer: the next generation. Cell **144**, 646–674 (2011)
135. Harrendorf, S., Heiskanen, M., Malby, S.: International Statistics on Crime and Justice. European Institute for Crime Prevention and Control, affiliated with the United Nations (2010)
136. Harsanyi, J.: Games with incomplete information played by "Bayesian" players, I-III part I. the basic model. Manag. Sci. **14**(3), 159–182 (1967)
137. Hartwell, H., Hopfield, J., Leibler, S., Murray, A.: From molecular to modular cell biology. Nature (1999)
138. Helbing, D.: Stochastic and Boltzmann-like models for behavioral changes, and their relation to game theory. Phys. A **193**(2), 241–258 (1993)
139. Helbing, D.: Traffic and related self-driven many-particle systems. Rev. Modern Phys. **73**, 1067–1141 (2001)
140. Helbing, D.: Quantitative Sociodynamics: Stochastic Methods and Models of Social Interaction Processes. Springer, Berlin (2010)
141. Helbing, D.: New ways to promote sustainability and social well-being in a complex, strongly interdependent world: the future ICT approach. Why Society is a Complex Matter. Lecture Notes in Mathematics (Ball, P.), pp. 55–60. Springer, Berlin (2012)
142. Helbing, D., Johansson, A.: Pedestrian crowd and evacuation dynamics. Encyclopedia of Complexity and System Science, pp. 6476–6495. Springer, Berlin (2009)
143. Helbing, D., Johansson, A., Al-Abideen, H.: Dynamics of crowd disasters: an empirical study. Phys. Rev. E **75**, 046109 (2007)
144. Helbing, D., Sigmeier, J., Lämmer, S.: Self-organized network flows. Netw. Heterog. Media **2**(2), 193–210 (2007)
145. Helbing, D., Yu, W.: The outbreak of cooperation among success-driven individuals under noisy conditions. Proc. Natl. Acad. Sci. **106**, 3680–3685 (2009)
146. Helbing, D., Yu, W.: The future of social experimenting. Proc. Natl. Acad. Sci. **107**(12), 5265–5266 (2010)
147. Hemelrijk, C., Hildenbrandt, H.: Self-organized shape and frontal density of fish schools. Ethology **114**, 245–254 (2008)
148. Henderson, L.: The statistics of crowd fluids. Nature **229**, 381–383 (1971)

149. Henrich, J., Boyd, R., Bowles, S., Camerer, C., Fehr, E., Gintis, H., McElreath, R.: In search of homo economicus: behavioral experiments in 15 small-scale societies. Am. Econ. Rev. **91**, 73–78 (2001)
150. Herrero, M.A.: Through a glass, darkly: biology seen from mathematics: comment on "toward a mathematical theory of living systems focusing on developmental biology and evolution: a review and perspectives" by N. Bellomo and B. Carbonaro. Phys. Life Rev. **8**(1), 21 (2011)
151. Hilbert, D.: Mathematical problems. Bull. Am. Math. Soc. **8**, 437–479 (1902)
152. Hildenbrandt, H., Carere, C., Hemelrijk, C.: Self-organized aerial displays of thousands of starlings: a model. Behav. Ecol. **21**, 1349–1359 (2010)
153. Hillen, T., Othmer, H.: The diffusion limit of transport equations derived from velocity-jump processes. SIAM J. Appl. Math. **61**, 751–775 (2000)
154. Hofbauer, J., Sigmund, K.: Evolutionary game theory. Bull. Am. Math. Soc. **40**(4), 479–519 (2003)
155. Hoogendoorn, S., Bovy, P.: Gas-kinetic modeling and simulation of pedestrian flows. Transp. Res. **1710**, 28–36 (2000)
156. Huges, R.: A continuum theory for the flow of pedestrians. Transp. Res. B **36**, 507–536 (2002)
157. Hughes, R.: The flow of human crowds. Annu. Rev. Fluid Mech. **35**, 169–182 (2003)
158. Ivan, M.S., Rogazinsky, S.V.: Theoretical analysis of traditional and modern schemes of the DSMC method. In: Beylich, A.E. (ed.) Proceedings of the 17th International Symposium on Rarefied Gas Dynamics, 1990, pp. 629–642 (1990)
159. Jäger, E., Segel, L.: On the distribution of dominance in populations of social organisms. SIAM J. Appl. Math. **52**(5), 1442–1468 (1992)
160. Jona Lasinio, G.: La matematica come linguaggio delle scienze della natura. Public Speech, Centro "E. De Giorgi", Scuola Normale Superiore, Pisa (2004)
161. Kirman, A.P., Zimmermann, J.B.: Economics with Heterogeneous Interacting Agents. Lecture Notes in Economics and Mathematical Systems., vol. 503. Springer, Berlin (2001)
162. Knopoff, D.: On the modeling of migration phenomena on small networks. Math. Models Methods Appl. Sci. **23**, 541–563 (2013)
163. Knopoff, D.: On a mathematical theory of complex systems on networks with application to opinion formation. Math. Models Methods Appl. Sci. **24**, 405–426 (2014)
164. Kogan, M.: Rarefied Gas Dynamics. Plenum Press, New York (1969)
165. Koura, K.: Null-collision technique in the direct simulation Monte Carlo technique. Phys. Fluids **29**, 3509–3511 (1986)
166. Kraus, M., Piff, P., Mendoza-Denton, R., Rheinschmidt, M., Keltner, D.: Social class, solipsism, and contextualism: how the rich are different from the poor. Psychol. Rev. **119**, 546–572 (2012)
167. Lasry, J.M., Lions, P.L.: Jeux champ moyen. I Le cas stationnaire. Comptes Rendus Mathematique **343**(10), 679–684 (2006)
168. Lasry, J.M., Lions, P.L.: Jeux champ moyen. II horizon fini et contrle optimal. Comptes Rendus Mathematique **343**(9), 619–625 (2006)
169. Lee, S., Pak, H., Chon, T.: Dynamics of prey-flock escaping behavior in response to predator's attack. J. Theor. Biol. **240**, 250–259 (2006)
170. LeVeque, R.: Finite Volume Methods for Hyperbolic Problems. Cambridge University Press, Cambridge (2002)
171. Longo, G., Preziosi, L.: On a conservative polar discretization of the Boltzmann equation. Jpn. J. Ind. Appl. Math. **14**, 399–435 (1997)
172. May, R.: Uses and abuses of mathematics in biology. Science **303**(5659), 790–793 (2004)
173. Mayr, E.: 80 years of watching the evolutionary scenario. In: Proceedings of the American Philosophical Society
174. Mayr, E.: The philosophical foundations of darwinism. Proc. Am. Philos. Soc. **145**, 488–495 (2001)
175. Mogilner, A., Edelstein-Keshet, L., Spiros, A.: Mutual interactions, potentials, and individual distance in a social aggregation. J. Math. Biol. **47**, 353–389 (2003)

176. Morgenstern, O., Von Neumann, J.: Theory of Games and Economic Behavior. Princeton University Press, Princeton (1953)
177. Motsch, S., Tadmor, E.: Heterophilious dynamics enhances consensus. SIAM Rev. **56**, 577–621 (2014)
178. Mouhot, C., Pareschi, L.: Fast algorithms for computing the Boltzmann collision operator. C. R. Acad. Sci. Paris **339**(1), 71–76 (2004)
179. Moussaïd, M., Helbing, D., Garnier, S., Johansson, A., Combe, M., Theraulaz, G.: Experimental study of the behavioural mechanisms underlying self-organization in human crowds. Proc. Roy. Soc. Ser. B **276**, 2755–2762 (2009)
180. Moussaïd, M., Theraulaz, G.: Comment les piétons marchent dans la foule. La Recherche **450**, 56–59 (2011)
181. Myatt, D.P., Wallace, C.: An evolutionary analysis of the volunteers dilemma. Games Econ. Behav. **62**, 67–76 (2008)
182. Nanbu, K.: Direct simulation scheme derived from the Boltzmann equation. I. monocomponent gases. Jpn. J. Phys. **19**, 2042–2049 (1980)
183. Nash, J.: Noncoperative games. Ann. Math. **54**, 287–295 (1951)
184. Newman, A., Barabási, L., Watts, D.: Dyn. Princeton University Press, Princeton, Process. Complex Netw (2006)
185. Nordsieck, A., Hicks, B.: Monte Carlo evaluation of the Boltzmann collision integral. In: Brudin, C.L. (ed.) Proceedings of the 9th International Symposium on Rarefied Gas Dynamics 1967, vol. 2, pp. 695–710 (1967)
186. Nowak, M.: Evolutionary Dynamics - Exploring the Equations of Life. Harvard University Press, Cambridge (2006)
187. Nowak, M.: Five rules for the evolution of cooperation. Science **314**(5805), 1560–1563 (2006)
188. Nowak, M., Ohtsuki, H.: Prevolutionary dynamics and the origin of evolution. Proc. Natl. Acad. Sci. **105**(39), 14924–14927 (2008)
189. Nowak, M., Sigmund, K.: Evolutionary dynamics of biological games. Science **303**(5659), 793–799 (2004)
190. Ohwada, T.: Structure of normal shock wave: direct numerical analysis of the Boltzmann equation for hard-sphere molecules. Phys. Fluids **5**(1), 217–234 (1992)
191. Okubo, A.: Dynamical aspects of animal grouping: swarms, schools, flocks, and herds. Adv. Biophys. **22**, 1–94 (1986)
192. Ormerod, P.: Crime: Economic Incentives and Social Networks. The Institute of Economic Affairs, London (2005)
193. Othmer, H., Dunbar, S., Alt, W.: Models of dispersal in biological systems. J. Math. Biol. **26**, 263–298 (1988)
194. Othmer, H., Hillen, T.: The diffusion limit of transport equations II: chemotaxis equations. SIAM J. Appl. Math. **62**, 1222–1250 (2002)
195. Outada, N., Vauchelet, N., Akrid, T., Khaladi, M.: From kinetic theory of multicellular systems to hyperbolic tissue equations: asymptotic limits and computing. Math. Models Methods Appl. Sci. **26**, 2709–2734 (2016)
196. Pareschi, L., Russo, G.: Numerical solution of the Boltzmann equation. I. spectrally accurate approximation of the collision operator. SIAM J. Numer. Anal. **37**(4), 1217–1245 (2000)
197. Pareschi, L., Toscani, G.: Interacting Multiagent Systems - Kinetic Equations and Monte Carlo methods. Oxford University Press, Oxford (2013)
198. Paveri Fontana, S.: On Boltzmann like treatments for traffic flow. Transp. Res. **9**, 225–235 (1975)
199. Perthame, B.: Transport Equations in Biology. Birkhäuser, New York (2007)
200. Peyton Young, H.: An evolutionary model of bargaining. J. Econ. Theory **59**, 145–168 (1993)
201. Platkowski, T., Illner, R.: Discrete velocity models of the Boltzmann equation: a survey on the mathematical aspects of the theory. SIAM Rev. **30**, 213–255 (1988)
202. Prigogine, I., Herman, R.: Kinetic Theory of Vehicular Traffic. Elsevier, New York (1971)
203. Pucci, P., Salvadori, M.: On an initial value problem modeling evolution and selection in living systems. Discret. Contin. Dyn. Syst. Ser. S **7**, 807–821 (2014)

204. Rand, D., Arbesman, S., Christakis, N.: Dynamic social networks promote cooperation in experiments with humans. Proc. Natl. Acad. Sci. **108**(48), 19193–19198 (2011)
205. Reed, R.: Why is mathematical biology so hard? Not. Am. Math. Soc. **51**, 338–342 (2004)
206. Rinaldi, S., Della Rossa, F., Dercole, F., Gragnani, A., Landi, P.: Modeling Love Dynamics. World Scientific, New York (2016)
207. Rjasanow, S., Wagner, W.: A stochastic weighted particle method for the Boltzmann equation. J. Comput. Phys
208. Rjasanow, S., Wagner, W.: Stochastic Numerics for the Boltzmann Equation. Springer, Berlin (2005)
209. Ronchi, E.: Disaster management: design buildings for rapid evacuation. Nature **528**, 333 (2015)
210. Ronchi, E., Nieto Uriz, F., Criel, X., Reilly, P.: Modelling large-scale evacuation of music festival. Fire Saf. **5**, 11–19 (2016)
211. Ronchi, E., Reneke, P., Peacock, R.: A conceptual fatigue-motivation model to represent pedestrian movement during stair evacuation. Appl. Math. Model. **40**, 4380–4396 (2016)
212. Rufus, I.: Differential Games: A Mathematical Theory with Applications to Warfare and Pursuit. Control and Optimization. Wiley, New York (1965)
213. Saint-Raymond, L.: Hydrodynamic Limits of the Boltzmann Equation. Lecture Notes in Mathematics, vol. 1971. Springer, Berlin (2009)
214. Santos, F., Pacheco, J., Lenaerts, T.: Evolutionary dynamics of social dilemmas in structured heterogeneous populations. Proc. Natl. Acad. Sci. **103**(9), 3490–3494 (2006)
215. Santos, F., Vasconcelos, V., Santos, M., Neves, P., Pacheco, J.: Evolutionary dynamics of climate change under collective-risk dilemmas. Math. Models Methods Appl. Sci. **22**, 1140004 (2012)
216. Schadschneider, A., Klingsch, W., Kläpfel, H., Kretz, T., Rogsch, C., Seyfried, A.: Evacuation dynamics: empirical results, modeling and applications. Encyclopedia of Complexity and System Science, pp. 3142–3176. Springer, Berlin (2009)
217. Schadschneider, A., Seyfried, A.: Empirical results for pedestrian dynamics and their implications for modeling. Netw. Heterog. Media **6**, 545–560 (2011)
218. Scheffer, M., Bascompte, J., Brock, W., Brovkin, V., Carpenter, S., Dakos, V., Held, H., van Nes, E., Rietkerk, M., Sugihara, G.: Early-warning signals for critical transitions. Nature **461**, 53–59 (2009)
219. Schelling, T.: Dynamic models of segregation. J. Math. Sociol. **1**, 143–186 (1971)
220. Schrödinger, E.: What is Life?. The Physical Aspect of the Living Cell. Cambridge University Press, Cambridge (1944)
221. Sigmund, K.: The Calculus of Selfishness. Princeton University Series in Theoretical and Computational Biology. Princeton University Press, Princeton (2011)
222. Simon, H.: Theories of decision-making in economics and behavioral science. Am. Econ. Rev. **49**, 253–283 (1959)
223. Simon, H.: Models of Bounded Rationality. MIT Press, Boston (1997)
224. Smith, M.J.: The stability of a dynamic model of traffic assignment - an application of a method of Lyapunov. Transp. Sci. **18**, 245–252 (1984)
225. Sone, Y., Ohwada, T., Aoki, K.: Temperature jump and Knudsen layer in a rarefied gas over a plane wall: numerical analysis of the linearized Boltzmann equation for hard-sphere molecules. Phys. Fluids **1**, 363–370 (1989)
226. Stadler, B., Stadler, P., Wagner, G., Fontana, W.: The topology of the possible: formal spaces underlying patterns of evolutionary change. J. Theor. Biol. **213**, 241–274 (2001)
227. Stefanov, S.K.: On DSMC calculations of rarefied gas flows with small number of particles in cells. SIAM J. Sci. Comput. **33**(2), 677–702 (2011)
228. Strogatz, S.: Exploring complex networks. Nature **410**(6825), 268–276 (2001)
229. Szabóa, G., Fáthb, G.: Evolutionary games on graphs. Phys. Report. **446**, 97–216 (2007)
230. Taleb, N.: The Black Swan: The Impact of the Highly Improbable. Random House, New York (2007)
231. Thaler, R.: From homo economicus to homo sapiens. J. Econ. Perspect. **14**, 133–141 (2000)

232. Thieme, H.: Mathematics in Population Biology. Princeton University Press, Princeton (2003)
233. Triebel, H.: The Structure of Functions. Birkhäuser, Basel (2001)
234. Triebel, H.: Theory of Function Spaces. Birkhäuser, Basel (2007)
235. Vega-Redondo, F.: Evolution, Games, and Economic Behaviour. Oxford University Press, Oxford (1996)
236. Vega-Redondo, F.: Complex Social Networks. Cambridge University Press, Cambridge (2007)
237. Vogelstein, B., Kinzler, K.: Cancer genes and the pathways they control. Nat. Med. **10**, 789–799 (2004)
238. Wagner, W.: A convergence proof of bird's direct simulation monte-carlo method for the Boltzmann equation. J. Stat. Phys. **66**, 1011–1044 (1992)
239. Wasserstein, L.: Markov processes over denumerable products of spaces describing large systems of automata. Probab. Inf. Trans. **5**, 47–52 (1969)
240. Webb, G.: Theory of Nonlinear Age-Dependent Population Dynamics. Dekker, New York (1985)
241. Weidlich, W., Haag, M.A.: Concepts and Models of a Quantitative Sociology The Dynamics of Interacting Populations. Springer, Berlin (1983)
242. Weinberg, R.: The Biology of Cancer. Garland Science - Taylor and Francis, New York (2007)
243. Wigner, E.: The unreasonable effectiveness of mathematics in the natural sciences. Commun. Pure Appl, Math (1960)
244. Wijermans, N.: Understanding Crowd Behaviour. University of Groningen, Ph.D. thesis (2011)
245. Wijermans, N., Conrado, C., van Steen, M., Martella, C., Li, J.: A landscape of crowd management support: an integrative approach. Saf. Sci. **86**, 142–164 (2016)
246. Wolpert, D.H.: Information theory - the bridge connecting bounded rational game theory and statistical physics. In: Braha, D., Minai, A.A., Bar-Yam, Y. (eds.) Complex Engineered Systems, Understanding Complex Systems, pp. 262–290. Springer, Princeton (2006)
247. Yakovenko, V.M., Barkley Rosser, J.: Statistical physics of social dynamics. Rev. Modern Phys. **4**, 1703–1725 (2009)
248. Yamori, Y.: Going with the flow: micro-macro dynamics in the macrobehavioral patterns of pedestrian crowds. Psychol. Rev. **105**, 530–557 (1998)
249. Yanitskiy, V.: Operator approach to direct simulation Monte Carlo theory in rarefied gas dynamics. In: Beylich, A.E. (ed.) Proceedings of the 17th International Symposium on Rarefied Gas Dynamics 1990, pp. 770–777 (1990)
250. Zhang, J.: A dynamic model of residential segregation. J. Math. Sociol. **28**, 147–170 (2004)

Index

N. Bellomo et al., *A Quest Towards a Mathematical Theory of Living Systems*, Modeling and Simulation in Science, Engineering and Technology,
DOI 10.1007/978-3-319-57436-3